BUILDING

ARCHITECTURAL ENGINEERING SURVEYING

# 建筑工程测量

中英双语版


主　编 | 黄　彬　张倩倩　丁于强

副主编 | 陈　乘　肖　莎　韦柳盛




湖南大學出版社·长沙

图书在版编目（CIP）数据

建筑工程测量：汉、英/黄彬，张倩倩，丁于强主编.—长沙：湖南大学出版社，2024.7

ISBN 978-7-5667-2960-6

Ⅰ.①建… Ⅱ.①黄… ②张… ③丁… Ⅲ.①建筑测量—高等职业教育—教材—汉、英 Ⅳ.①TU198

中国国家版本馆CIP 数据核字（2023）第 075522号

**建筑工程测量（中英双语版）**

JIANZHU GONGCHENG CELIANG（ZHONG-YING SHUANGYU BAN）

**主　　编**：黄　彬　张倩倩　丁于强
**策划编辑**：贾志萍　胡建华
**责任编辑**：尹鹏凯　贾志萍
**印　　装**：长沙市雅捷印务有限公司
**开　　本**：787 mm × 1092 mm　1/16　**印　　张**：12.75　**字　　数**：343千字
**版　　次**：2024年7月第1版　**印　　次**：2024年7月第1次印刷
**书　　号**：ISBN 978-7-5667-2960-6
**定　　价**：58.00元

**出 版 人**：李文邦
**出版发行**：湖南大学出版社
**社　　址**：湖南 · 长沙 · 岳麓山　**邮　　编**：410082
**电　　话**：0731-88822559（营销部）　88821174（编辑部）　88821006（出版部）
**传　　真**：0731-88822264（总编室）
**网　　址**：http://press.hnu.edu.cn

# 目　录

Contents

3



4

# 7

建筑工程测量
ARCHITECTURAL ENGINEERING SURVEYING

# 1

# 测量的基本知识

## Basic Knowledge of Surveying

**学习目标：**

**Learning Objectives:**

通过本章的学习，了解测量学的定义及分类、地面点的坐标、空间直角坐标系、用水平面代替水准面的范围、测量误差的产生原因及分类；熟悉建筑工程测量的任务和作用、测量工作的基准面和基准线、测量工作的基本原则、衡量测量精度的指标、误差传播定律；掌握确定地面点位和高程的方法。

Through the study of this chapiter, to learn about the definition and classification of surveying, the coordinates of ground points, the spatial rectangular coordinate system, the range of replacing a level surface with a horizontal plane, and the causes and classification of observation errors; to get familiar with the tasks and functions of architectural engineering surveying, the datum plane and datum line for surveying, the basic principle of surveying, the indicators to measure observation accuracy, and the law of error propagation; to master the method of determining ground points and elevations.

**能力目标：**

**Ability Objectives:**

能够确定地面点的平面位置和高程位置，能够根据测量工作的基本程序和原则实施测量工作。

Be able to determine the plane position and elevation position of ground points, and carry out surveys according to the basic procedures and principles of surveying.

# 1.1 建筑工程测量概述

# 1.1 Overview of Architectural Engineering Surveying

## 1.1.1 测量学

## 1.1.1 Surveying

### （1）测量学的发展历史

### (1) Development history of surveying

测量学是一门历史悠久的科学，早在几千年前，我国古代人民已把测量技术应用于土地划分、河道整治及地域图测绘等领域。在我国，追溯到上古时代，已有夏禹在黄河两岸治理水患的传说。这些活动都需要一定的测量知识及简单的测量工具的参与。

Surveying is a branch of science with a long history. Thousands of years ago, the ancient Chinese people have applied surveying technology to land division, river regulation, and surveying and mapping of native maps. In China, there are legends of King Yu of the Xia Dynasty about taming the floods on both sides of the Yellow River, revealing the initial application of some surveying knowledge or the beginning stage of surveys with simple tools.

测量学是一门与时俱进、蓬勃发展的科学。测量学一开始被用于土地整理，随着社会生产的发展，它逐渐被应用到社会的许多生产部门。早在战国时期，我国人民就发明了指南针（图1-1），之后又创制了浑天仪等测量仪器（图1-2），并绘制了相当精确的全国地图。17世纪初，经纬仪开始应用于天象观测（图1-3），这是测绘科学发展史上所发生的较大变革。自此以后，经纬仪普遍应用于各种测量与观测活动。19世纪末，航空摄影测量的发展，为测量学增添了新的内容。20世纪40年代，自动安平水准仪的问世，标志着水准测量自动化的开端。20世纪50年代前后，电子学、信息论、相干光理论、电子计算技术、空间科学技术等新的科学理论与技术的迅速发展推动了测绘科学的发展。20世纪60年代，由于电子计算技术的飞速发展，自动化程度很高的电子经纬仪、电子全站仪和自动绘图仪出现了。20世纪70年代，人造卫星应用黑白、单光谱段、多光谱段及彩色红外等摄影方式拍摄地球的照片，使航天技术有了广泛的应用与发展。卫星运行的高度比飞机高几十倍到几百倍，故视野宽广，覆盖面积大，可以对同一地区重复摄影，便于监测自然现象的变化，并且不受地理及气候条件的限制，有利于对深山、荒漠及海洋进行勘测。20世纪80年代，利用电磁波测距仪进行的距离测量，其误差可精确到厘米级。20世纪90年代，全球定位系统卫星成功发射，只要在地面欲测点安置接收卫星信号的测量设备，就可以很快地确定地面点的位置，相距1 km的两点之间的相对精度可达5~10 mm。随着现代科学技术的发展，测量科学也必将向更高层次的自动化方向和数字化方向发展。

Surveying is a flourishing branch of science keeping pace with the times. Initially used for land consolidation, it has been gradually applied to many sectors of production, with the development of human society. As early as the Warring States Period, the ancient people of China invented the compass(Figure 1-1), created measuring instruments such as the armillary sphere(Figure 1-2), and made fairly accurate national maps. At the beginning of the 17th century, theodolite began to be used for astronomical observation(Figure 1-3), marking a great change in the history of surveying and mapping science. Since then, theodolite have been widely used for various measurements and observations. At the end of the 19th century, aerial photogrammetry was developed into a new branch subject of surveying. In the 1940s, the introduction of automatic compensated levels marked the beginning of leveling automation. Around the 1950s, the rapid development of new science and technology fields such as electronics, information theory, partially coherent light theory, electronic computing technology, and space science and technology accelerated the development of surveying and mapping science. In the 1960s, due to the rapid development of electronic computing technology, electronic theodolites, electronic total stations and auto-plotters with a high degree of automation emerged. In the 1970s, satellite-based black-and-white, single-spectral, multi-spectral, and color infrared photography to take photos of the earth expanded the application and development scopes of aerospace technology. From an altitude of dozens to hundreds of times higher than that of aircrafts, satellites have a wide field of view and a large coverage area, enabling repeated photographic shooting in the same area, thus creating convenient conditions for monitoring changes in natural phenomena, without any geographical and climatic restrictions, so satellite photography is conducive to surveys to remote mountains, deserts and oceans. In the 1980s, distance measurements were conducted with electromagnetic distance measuring instruments with errors of only centimeters. In the 1990s, depending on the GPS satellites successfully launched, any ground point could be quickly positioned by the surveying equipment placed at such a point for receiving satellite signals, with a relative accuracy of 5-10 mm between two points 1 km apart. With the development of modern science and technology, surveying science is bound to develop toward higher-level automation and digitalization.

中华人民共和国成立以后，我国测量事业进入了一个崭新的发展阶段。1956 年，国家测绘总局成立，1957 年，中国科学院测量与地球物理研究所成立。此后各业务部门也纷纷设立测绘机构，培养测绘人员的各级学校也相继建立。随着测绘队伍飞速壮大，测绘科学技术也得到快速发展。我国已建成了全国绝大部分地区的大地控制网，近年来还建立了全国的 GPS 控制网，测量完成了大量不同比例尺的基本地形图，各种工程建设的测量工作也取得了显著成绩。仪器制造方面，我国从无到有，现在不仅能够生产系列的光学测量仪器，还成功研制了各种测程的光电测距仪、卫星激光测距仪、解析测图仪、激光垂准仪、激光扫平仪和全站仪等先进仪器。在测量技术方面，我国紧紧跟随世界最新测绘技术发展的脚步，进入了测绘现代化的高速发展阶段。

After the founding of the People's Republic of China, China's surveying industry ushered in a new stage of development. In 1956, the State Bureau of Surveying and Mapping was established. In 1957, the Institute of Geodesy and Geophysics of the Chinese Academy of Sciences was founded. After that, surveying and mapping institutions were set up in various industrial sectors, and college departments were established to foster surveying and mapping technicians. The rapidly growing pool of talents laid a solid foundation for the fast development of surveying and mapping science. China has built a geodetic control network in most areas across the country and has also established a national GPS control network in recent years. A large number of basic topographic maps on different scales have been drawn on the basis of massive survey data, and remarkable progress in surveys has been made for engineering construction. From the perspective of the development of instrument manufacturing, China has successfully developed leading-edge instruments from the scratch, such as electro-optical distance measuring instruments, satellite laser rangers, analytical plotters, laser plummets, laser levelers and total stations with varied rangings, in addition to a series of optical measuring instruments. From the perspective of surveying technology, keeping a close eye on the global latest surveying and mapping technology, China has been experiencing a rapid development stage of modern surveying and mapping.

1992 年，我国政府制定了新中国第一部《中华人民共和国测绘法》，1993 年 7 月 1 日起施行；2002 年又公布了修订后的《中华人民共和国测绘法》，自 2002 年 12 月 1 日起施行，这标志着我国测量工作进入依法工作、规范发展的新阶段。

In 1992, the Chinese government adopted the Surveying and Mapping Law of the People's Republic of China, which came into effect on July 1, 1993. In 2002, the Surveying and Mapping Law of the People's Republic of China (Revision) was released, which came into effect on December 1st, 2002, opening a new stage of legal and standardized development of the surveying sector in China.

图1-1 司南

Figure 1-1 Compass

图1-2 浑天仪

Figure 1-2 Armillary Sphere

图1-3 经纬仪

Figure 1-3 Theodolite

## （2）测量学的分类

### (2) Classification of surveying

测量学是一门内容体系比较庞大的学科。根据研究对象、应用范围和测量手段等的不同，测量学通常可分为以下几个分支学科。

As a discipline with a vast system of knowledge, surveying science is divided into the following branch disciplines according to different research objects, application scopes and surveying means.

①普通测量学。

① Elementary surveying.

普通测量学是研究将地球自然表面局部地区的地物和地貌按一定比例尺测绘成大比例尺地形图的基本理论和方法的学科，属测量学的基础部分。

As the fundamental part of this discipline, elementary surveying involves the basic theories and methods of surveying and mapping the ground objects and land features of local areas on the natural surface of the earth into large-scale topographic maps.

②大地测量学。

② Geodesy.

大地测量学是研究地球形状、大小和重力场及其变化，通过建立区域和全球三维控制网、重力网及利用卫星测量等方法测定地球各种动态的理论和技术的学科。

Targeting the shape, size and gravity field of the earth and their changes, geodesy involves the theories and techniques to measure various dynamics of the earth by means of establishing regional and global 3D control networks, gravity networks, and satellite surveying, etc.

③摄影测量学。

③ Photogrammetry.

摄影测量学是研究利用摄影或遥感技术获取地物和地貌的影像，并进行分析处理，以绘制地形图或获得数字化信息的理论和方法的学科。其中，航空摄影测量是测绘中、小比例尺国家基本地形图的主要方法，现已应用到大比例尺地形图的测绘中；而近景摄影测量已经在古建筑测绘、建（构）筑物的变形观测、动态目标测量等许多方面得到了广泛的应用。

Photogrammetry involves the theories and methods to acquire images of ground objects and land features by using photographic or remote sensing techniques and to produce topographic maps or to generate digital information after analysis and processing of these images. Aerial photogrammetry is mainly used for surveying and mapping of national basic topographic maps on a medium or small scale, and has been applied to the surveying and mapping of large-scale topographic maps; close-range photogrammetry has been widely applied in many aspects such as surveying and mapping of ancient buildings, deformation observation of buildings (structures), and dynamic target measurement.

④工程测量学。

④ Engineering surveying.

工程测量学是研究工程建设和自然资源开发中各个阶段进行的控制测量、地形测绘、施工放样、变形监测及建立相应信息系统的理论和技术的学科。其主要内容包括测绘满足工程规划和勘察设计需要的大比例尺地形图；将图纸上设计的建（构）筑物轴线桩位标定到地面上；对在施工过程中及竣工后建(构)筑物的变形进行监测。

Engineering surveying involves the theories and techniques of control survey, topographic survey and mapping, construction setting-out, deformation monitoring, and establishing corresponding information systems in various stages of engineering construction and natural resource exploitation. This branch discipline mainly includes surveying and mapping of large-scale topographic maps to meet the requirements of planning, survey and design of engineering works; stake marking of the axis of buildings (structures) designed on the drawings to the ground; deformation monitoring of buildings (structures) during construction and after completion.

⑤海洋测绘学。

⑤ Marine surveying and mapping.

海洋测绘学是研究海洋定位，测定海洋大地水准面和平均海面、海底和海面地形、海洋重力、磁力、海洋环境等自然和社会信息的地理分布，以及编制各种海图的理论和技术的学科。其主要内容包括海洋大地测量、水深测量、海底地形测量、海洋重力测量、海岸地形测量、海道测量、海洋专题测量和海图测绘等。

Marine surveying and mapping involves the theories and techniques of the natural and social information study from the view of geographical distribution, such as marine positioning, measurement of ocean geoid and mean sea level, seafloor and sea surface topography, ocean gravity, magnetism, marine environment, as well as compiling various nautical charts. This branch discipline mainly includes marine geodetic surveying, sounding surveying, seafloor topographic surveying, marine gravity surveying, coast topographic surveying, hydrographic surveying, marine thematic surveying and nautical chart mapping.

⑥地图制图学。

⑥ Cartography.

地图制图学是研究各种地图的制作理论、原理、工艺技术和应用的一门学科。其主要内容包括地图编制、地图投影学、地图整饰、地图印刷等。现代地图制图学正向着制图自动化、电子地图制作及地理信息系统方向发展。

Cartography involves the theories, principles, techniques, and applications of map-making for various purposes. This branch discipline mainly includes map compilation, map projection, map appearance, and map printing. Modern cartography is developing towards automatic and electronic map-making and GIS fields.

### （3）测量学的基本任务

### (3) Basic tasks of surveying

对工程建设而言，测量学的基本任务包括测定和测设两个方面。

For the purpose of engineering construction, the basic tasks of surveying include measurement and setting-out.

①测定。

① Measurement.

测定是指使用测量仪器和工具，按照测量的有关原理和方法，将地球表面的地物和地貌绘制成地形图，为经济建设、国防建设和科学研究等服务。

Measurement is to draw topographic maps of ground objects and land features of the earth surface with measuring instruments and tools, and according to the relevant surveying principles and methods, so as to serve economic construction, national defense development and scientific research .

②测设。

② Setting-out.

测设是指使用测量仪器和工具，按照测量的有关原理和方法，将图纸上规划设计好的建（构）筑物的平面位置和高程，在实地标定出来，作为施工的依据，故又称为施工放样，它是工程设计与工程施工之间的桥梁。

Setting-out refers to field making of the plane position and elevation of buildings (structures) planned and designed on the drawings with the measuring instruments and tools, and according to the relevant surveying principles and methods, as the basis for construction, which is also called construction setting-out, and serves as a bridge linking engineering design and engineering construction.

## 1.1.2 建筑工程测量的任务

## 1.1.2 Tasks of Architectural Engineering Surveying

建筑工程测量属于工程测量学范畴，它指建筑工程在勘察设计、施工建设和组织管理等阶段，应用测量仪器和工具，采用一定的测量技术和方法，根据工程施工进度和质量要求，完成应进行的各种测量工作。建筑工程测量的主要任务如下。

Falling into the category of engineering surveying, architectural engineering surveying is to carry out various surveys according to the construction schedule and quality requirements of the engineering construction by using measuring instruments and tools in the stages of survey and design, construction and organization management of building works. The main tasks of architectural engineering surveying are as follows.

### (1) 大比例尺地形图的测绘

### (1) Surveying and mapping of large-scale topographic maps

把工程建设区域内的各种地面物体的位置、性质以及地面的起伏形态，依据规定的符号和比例尺绘制成地形图，为工程建设的规划设计提供需要的图纸和资料。

The location and nature of various ground objects and the surface relief within the engineering construction area are drawn into topographic maps according to the specified symbols and scales to provide drawings and data for engineering construction up to the planning and design requirements.

### (2) 施工放样和竣工测量

### (2) Construction setting-out and acceptance survey

把图上设计的建（构）筑物按照设计的位置在实地标定出来，作为施工的依据；配合建筑施工，进行各种测量工作，保证施工质量；开展竣工测量，为工程验收、日后扩建和维修管理提供资料。

The buildings (structures) designed on the drawings are marked on the spot according to the design position as the basis for construction; various surveys are carried out in coordination with the engineering construction, to ensure the construction quality; the acceptance survey is carried out to provide data for project acceptance, future expansion and maintenance management.

### (3) 建（构）筑物的变形观测

### (3) Deformation observation of buildings (structures)

对一些大型的、重要的或位于不良地基上的建（构）筑物，为了在施工运营期间确保安全，需要定期对其进行变形观测以检测其稳定性。观测结果可作为对建筑设计、地基、材料、施工方法等的验证依据，起到提供基础研究资料的作用。

For some large and important buildings (structures) or those built on unfavorable ground, in order to ensure safety during construction and operation, it is necessary to carry out regular deformation observation for the purpose of stability tracking. The deformation observation results can also be used as a verification basis for building's design, ground treatment, selection of materials and construction methods, etc., and play a role in providing data for basic research.

## 1.1.3 建筑工程测量的作用

## 1.1.3 Functions of Architectural Engineering Surveying

建筑工程测量在工程建设中有着广泛的应用，它服务于工程建设的每一个阶段。

Architectural engineering surveying is widely applied and serves every stage in engineering construction.

①在工程勘测阶段，测绘地形图为规划设计提供各种比例尺的地形图和测绘资料。

① In the engineering survey stage, the topographic maps on various scales and data are generated from surveying and mapping for planning and design purposes.

②在工程设计阶段，应用地形图进行总体规划和设计。

② In the engineering design stage, the topographic maps are used for overall planning and design.

③在工程施工阶段，要将图纸上设计好的建筑物、构筑物的平面位置和高程按设计要求测设于实地，以此作为施工的依据；在施工过程中用于土方开挖、基础和主体工程的施工测量；在施工中还要经常对施工和安装工作进行检验、校核，以保证所建工程符合设计要求；

工程竣工后，还要进行竣工测量，绘制施测竣工图，供日后扩建和维修之用。

③ In the engineering construction stage, the plane position and elevation of buildings and structures designed on the drawings are set out on the site according to the design requirements, as the basis for construction; construction surveys are conducted for earth excavation, foundation and main works during construction; during the construction, the construction and installation works are inspected and checked frequently to ensure their compliance to the design requirements; after the completion, acceptance surveys are conducted to produce as-built drawings as a reference for future expansion and maintenance.

④在工程管理阶段，对建筑物和构筑物进行变形观测，以保证工程的安全性。

④ In the engineering management stage, deformation observation is kept on buildings and structures, to ensure construction safety.

总而言之，在工程建设的各个阶段都需要进行测量工作，并且测量的精度和速度直接影响到整个工程的质量和进度。

To sum up, surveys are necessary at all stages of engineering construction, and survey accuracy and rate directly affect the quality and progress of the entire engineering construction.

# 1.2　地面点位的确定

## 1.2 Determination of Ground Points

地球表面上的点称为地面点。测量工作的实质就是确定地面点的空间位置。因为地球表面上的地物和地貌的形状可认为是由点、线、面构成的，其中，点是最基本的单元，所以，地面点位的确定是测量工作中最基本的问题。

Points on the earth surface are regarded as ground points. The essence of surveying is to determine the spatial position of ground points. Since the configurations of ground objects and land features on the earth surface can be considered to be composed of points, lines and planes, among which points are the most basic units, the determination of ground points is the most basic task in surveys.

### 1.2.1 地球的形状和大小

### 1.2.1 Shape and Size of the Earth

测量工作的主要研究对象是地球的自然表面。地球的自然表面极为复杂，有山地、丘陵、平原、盆地、高原等高低起伏的地形。其中，珠穆朗玛峰海拔 8 848.86 m，马里亚纳海沟深度约 11 034 m。但是这样的高低起伏，相对于地球巨大的半径来说还是很小的，仍可忽略不计。地球表面上海洋的面积约占 71%，而陆地面积约占 29%，因此，人们设想有一个静止的海水面，向陆地延伸并包围整个地球，形成一个封闭的曲面，将这个曲面看作地球的形体。

The main research object of surveys is the natural surface of the earth, which is extremely complex and undulated along with mountains, hills, plains, basins, and plateaus. The Mount Qomolangma is 8 848.86 m above the sea level, while the Mariana Trench is 11 034 m below the sea level. However, such ups and downs are still negligible relative to the huge radius of the earth. Since the oceans occupy about 71 percent of the surface area of the earth and the land area takes up only 29 percent, a still sea surface is envisaged to extend until the entire earth is encompassed into a closed curved surface that is regarded as the shape of the earth.

由于潮汐的作用，海水面高低不同。假定其中有一个平均高度的静止海水面，则它所包围的形体称为大地体，代表了地球的形状与大小。我们将这个平均高度的静止海水面称为大地水准面。大地水准面上的重力位处处相等，并与其上的重力方向处处保持着正交。地球上任何一点都要受到地球引力和地球自转引起的离心力的作用，这两个力的合力称为重力。重力的方向线称为铅垂线，所以水准面处处与铅垂线正交。铅垂线是测量工作的基准线，大地水准面是测量工作的基准面。

Sea surface varies at levels under the action of tides. Assuming that there is a still sea surface at the mean sea level, the shape of such a surface envelope is called the geoidal body, representing the shape and size of the earth. This still sea surface at the mean sea level is called the geoid. The gravitational potential on the geoid is equal everywhere and orthogonal to the direction of gravity acting on it. Any point on the earth is subject to earth attraction and centrifugal force caused by the rotation of the earth, the resultant of which is called gravity. The direction line of gravity is called the plumb line, so the geoid is orthogonal to the plumb line everywhere. The plumb line is used as the datum line and the geoid as the datum plane in surveys.

地球内部质量分布不均匀，引起铅垂线的方向产生不规则的变化，致使大地水准面成为一个不规则的复杂曲面，所以无法在这个曲面上进行测量数据的处理。为了使用方便，通常用一个非常接近于大地水准面，并可用数学式表示的几何形体（即地球椭球）来代替地球的形状作为测量计算工作的基准面，如图 1-4 所示。

Due to the non-uniform distribution of mass inside the earth, the direction of the plumb line changes irregularly, making the geoid an irregular and complex curved surface, and it is impossible to process the surveying data based on it. For the sake of convenience, a geometric figure that is very approximate to the geoid and can be expressed mathematically (i.e., the earth ellipsoid) is usually used to replace the geoid as the datum plane for surveys and computations, as shown in Figure 1-4 .

地球椭球是一个椭圆绕其短轴旋转而成的形体，故地球椭球又称旋转椭球。旋转椭球体

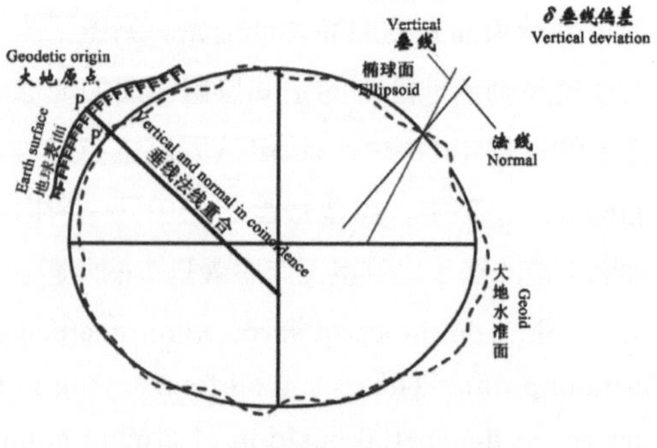


图1-4　大地水准面与地球椭球面

Figure 1-4 Geoid and Earth Ellipsoid

由长半轴 $a$ 和短半轴 $b$ 或扁率 $f$ 决定，如图 1-5 所示。

The earth ellipsoid is a figure created by rotating an ellipsoid around its short axis, so the earth ellipsoid is also called a rotating ellipsoid. The rotating ellipsoid is determined by a long radius indicated as $a$ and a short radius as $b$, or oblateness as $f$（Figure 1-5）.

目前，我国采用的 2000 国家大地坐标系，是我国当前最新的国家大地坐标系，英文缩写为 CGCS2000。2000 国家大地坐标系是全球地心坐标系在我国的具体体现，其原点为包括海洋和大气的整个地球的质量中心。$Z$ 轴指向 BIH1984.0 定义的协议极地方向 (BIH 国际时间局)，$X$ 轴指向 BIH1984.0 定义的零子午面与协议赤道的交点，$Y$ 轴按右手坐标系确定。2000 国家大地坐标系采用的地球椭球参数如下。

The China Geodetic Coordinate System 2000 is currently adopted in China as the latest one and is abbreviated as CGCS2000. CGCS2000 is the concrete embodiment of the global geocentric coordinate system in China, and its origin is the mass center of the entire earth including oceans and atmosphere. The $Z$-axis points to the protocol

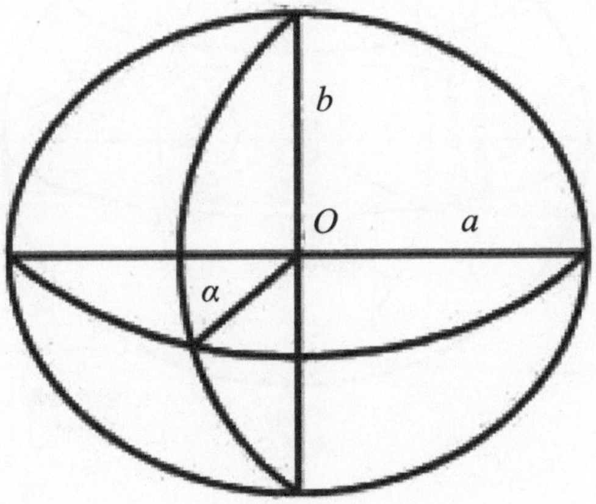


图1-5　参考椭球面

Figure 1-5 Reference Ellipsoid

polar direction defined by BIH1984.0 (BIH International Time Bureau), the $X$-axis points to the intersection of the zero meridian plane defined by BIH1984.0 and the protocol equator, and the $Y$-axis is determined by the right-hand coordinate system. The earth ellipsoid parameters used in CGCS2000 are as follows.

长半轴 $a$=6 378 137 m

Major semi-axis $a$=6 378 137 m

扁率 $f$=1/298.257222101

Oblateness $f$=1/298.257222101

地心引力常数 $GM$=3.986004418 × 1014 $m^3 \cdot s^{-2}$

Geocentric gravitational constant

$GM$=3.986004418 × 1014 $m^3 \cdot s^{-2}$

自转角速度 $\omega$=7.292115 × $10^{-5}$ $rad \cdot s^{-1}$

Rotational angular velocity $\omega$=7.292115 × $10^{-5}$ $rad \cdot s^{-1}$

短半轴 $b$=6 356 752.31414 m

Minor semi-axis $b$=6 356 752.31414 m

### 1.2.2 地面点位置的坐标系

### 1.2.2 Coordinate System of Ground Point Location

为了确定地面点的位置，需要建立坐标系。在测量工作中，可用地理坐标系、平面直角坐标系和空间直角坐标系表示地面点的位置。

In order to determine the position of ground points, a coordinate system needs to be established. The position of ground points in surveys can be represented in a geographical coordinate system, the plane rectangular coordinate system, and the spatial rectangular coordinate system.

#### （1）地理坐标系

#### (1) Geographical coordinate system

地理坐标系是用经纬度表示地面点位置的球面坐标，可分为天文坐标系和大地坐标系。

A geographical coordinate system is a

spherical coordinate system to represent the position of ground points with latitude and longitude values and is either an astronomical coordinate system or a geodetic coordinate system.

①天文坐标系。

① Astronomical coordinate system.

天文坐标系用于表示地面点在大地水准面上的位置，其基准是铅垂线和大地水准面，它用天文经度 $\lambda$ 和天文纬度 $\varphi$ 两个参数来表示地面点在球面上的位置。

An astronomical coordinate system is used to represent the position of ground points on the geoid. With the datum composed of the plumb line and the geoid, it represents the position of ground points on the sphere with the astronomical longitude $\lambda$ and latitude $\varphi$.

过地面上任一点 $P$ 的铅垂线与地球的旋转轴 NS 所组成的平面称为该点的天文子午面。天文子午面与大地水准面的交线称为天文子午线（也称经线），如图 1-6 所示。

The plane formed by the plumb line passing through any point $P$ on the ground and the rotation axis NS of the earth is called the astronomical meridian plane of this point. The intersection of the astronomical meridian plane and the geoid is called the astronomical meridian (also known as the longitude), as shown in Figure 1-6 .

设 $G$ 点为英国格林尼治天文台的位置，过 $G$ 点的天文子午面称为首子午面。

Point $G$ is the location of the Royal Greenwich Observatory in the United Kingdom, and the astronomical meridian plane passing through point $G$ is called the prime meridian plane.

$P$ 点天文经度 $\lambda$ 的定义是：过 $P$ 点的天文子午面 $NPKS$ 与首子午面 $NGMS$ 的两面角。从首子午线（本初子午线）向东或向西计算，取值范围为 0° ~ 180° 。在首子午线以东者为东经；以西者为西经。同一子午线上各点的经度相同。过 $P$ 点垂直于地球旋转轴的平面与地球表面的交线称为 $P$ 点的纬线，其所在平面过球心的纬线称为赤道。

The astronomical longitude $\lambda$ at point $P$ is defined as the angle between the astronomical meridian plane $NPKS$ passing through point $P$ and the prime meridian plane $NGMS$. From the prime meridian to the east or west in a range of 0°-180°, the zone in the east is represented as the east longitude; the zone in the west is represented as the west longitude. The points on the same meridian are represented with the same longitude. The intersection of the plane passing through point $P$ and perpendicular to the earth's axis of rotation and the earth's surface is called the latitude of point $P$, and the latitude of the plane passing through the center of the sphere is called the equator.

②大地坐标系。

② Geodetic coordinate system.

大地坐标系是表示地面点在参考椭球面上的位置，其基准是法线和参考椭球面，它用大

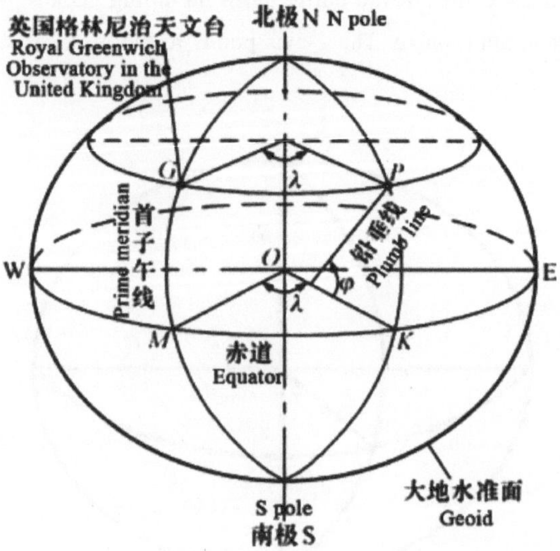


图1-6　天文坐标系

Figure 1-6 Astronomical Coordinate System

地经度 $L$ 和大地纬度 $B$ 表示。

A geodetic coordinate system represents the position of ground points on the reference ellipsoid. With the datum composed of the normal and the reference ellipsoid, it is represented by the geodetic longitude $L$ and the geodetic latitude $B$.

地面点 $P$ 的经度是指过该点的子午面与首子午线之间的夹角，用 $L$ 表示。经度从首子午线起算，往东自 0° 至 180° 经线称为东经；往西自 0° 至 180° 经线称为西经。地面点 $P$ 的纬度是指过该点的法线与赤道面之间的夹角，用 $B$ 表示。纬度从赤道面起算，往北自 0° 至 90° 纬线称为北纬；往南自 0° 至 90° 纬线称为南纬。我国地处北半球，各地的纬度都是北纬（图 1-7）。

The longitude of the ground point $P$ refers to the angle between the meridian plane passing through the point and the prime meridian, represented by $L$. The zone in 0° to 180° eastward from the prime meridian is called east longitude and west longitude is from 0° to 180° westward. The latitude of the ground point $P$ refers to the angle between the normal passing through the point and the equatorial plane, represented by $B$. The zone in 0° to 90° northward from the equatorial plane is called north latitude and south latitude is from 0° to 90° southward. China is located in the Northern Hemisphere and is represented by north latitudes (Figure 1-7).

## （2）平面直角坐标系

(2) Plane rectangular coordinate system

为了在工程测量中使用方便，常采用平面直角坐标系来表示地面点位，下面主要介绍常用的平面直角坐标系——高斯平面直角坐标系。

For the sake of convenience in engineering surveying, a plane rectangular coordinate system is often used to represent ground points. A common used plane rectangular coordinate system

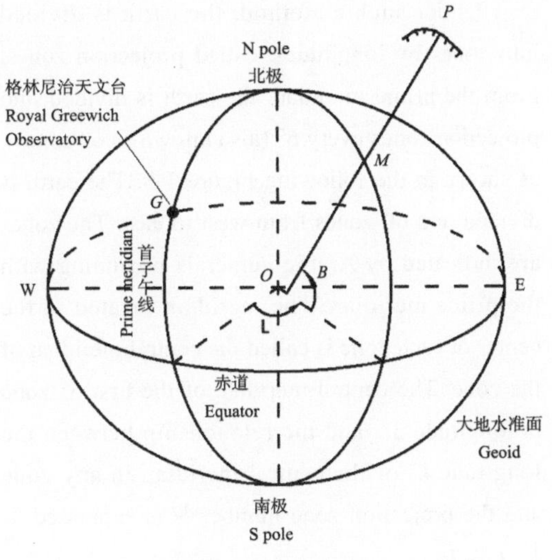


图1-7　大地坐标系

Figure 1-7 Geodetic Coordinate System

is introduced below—Gauss plane rectangular coordinate system.

当测量范围较大时，大地水准面不能再看成平面，而应作为椭球面处理。球面上不能建立直角坐标系。因此，可采用投影的方法将球面变为平面，然后再建立平面直角坐标系。我国采用的是高斯投影法。

The geoid is regarded as an ellipsoid for surveys in a large range, instead of a plane. Since a rectangular coordinate system cannot be established on a sphere, the sphere is converted into a plane by the projection method, and then a plane rectangular coordinate system is established. The Gauss projection method is commonly adopted in China.

高斯投影法首先将地球按经线划分成带，称为投影带。投影带是从首子午线起，每隔经度 6° 划分为一带（称为 6° 带），如图 1-8 所示。自西向东将整个地球划分为 60 个带。带号从首子午线开始，用阿拉伯数字表示，位于各带中央的子午线称为该带的中央子午线。第一个 6° 带的中央子午线的经度为 3°，任意一个带中央子午线经度 $L_0$ 与投影带号 $N$ 的关系为 $L_0=6N-3$。

Under such a method, the earth is divided into zones by longitudes, called projection zones. From the prime meridian, the earth is divided into projection zones every 6° (also known as 6° zones), as shown in the following Figure 1-8. The earth is divided into 60 zones from west to east. The zones are indicated by Arabic numerals beginning with the prime meridian. The meridian located in the center of each zone is called the central meridian of the zone. The central meridian of the first 6° zone is longitude 3°, and the relationship between the longitude $L_0$ of the central meridian in any zone and the projection zone number $N$ is expressed as $L_0=6N-3$.

反之，已知地面任一点的经度 $L$，则该点所在的 6° 带编号的公式为

Reversely, if the longitude $L$ of any point on the ground is known, the 6° zone number is expressed in the formula

$$N=\text{Int}[(L+3)/6+0.5] \quad (1.1)$$

式中，Int 为取整函数。

Where, Int is a rounding function.

地球分带如图 1-9 所示。投影后的经纬线图形有以下特点：

Earth zoning is shown in Figure 1-9. The projected longitudes and latitudes are characterized as follows:

(a) 投影后的中央子午线为直线，无长度变化。其余的经线投影为凹向中央子午线的对称曲线，长度较球面上的相应经线略长。

(a) The central meridian projected is a straight line without any length change. The remaining longitudes projected are symmetrical curves concave towards the central meridian, slightly longer than the corresponding longitudes on the sphere.

(b) 赤道的投影也为一直线，并与中央子午线正交。其余的纬线投影为凸向赤道的对称曲线。

(b) The projected equator is also a straight line orthogonal to the central meridian. Other latitudes projected are symmetrical curves convex towards the equator.

(c) 经纬线投影后仍然保持相互垂直的关系，说明投影后的角度无变形。

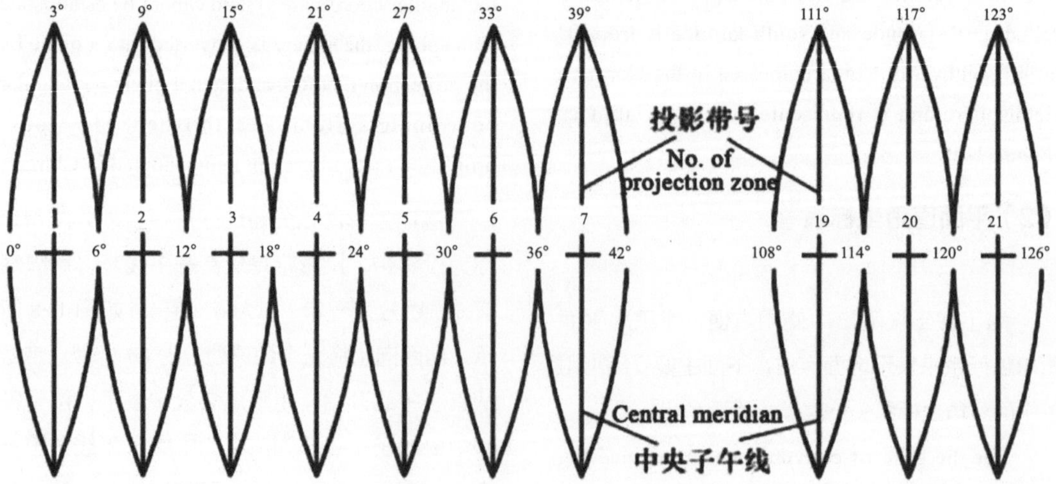


图1-8　6° 带中央子午线及带号

Figure 1-8 Central Meridian and Zone Number of 6° Zones

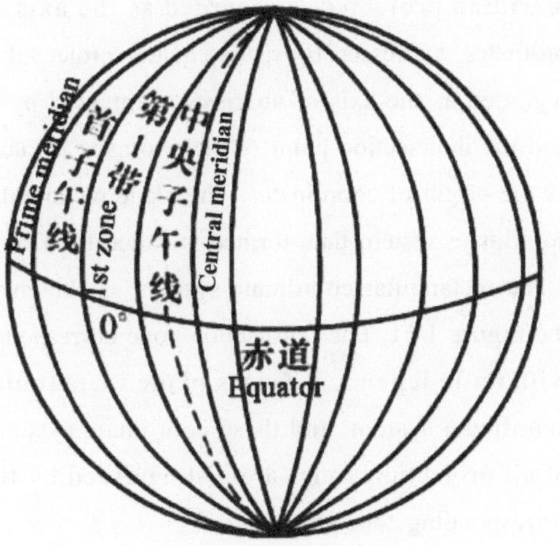


图1-9　地球分带

Figure 1-9 Earth Zoning

(c)The projected longitudes and latitudes are still perpendicular to each other, indicating no angular deformation after projection.

高斯投影没有角度变形，但有长度变形和面积变形，离中央子午线越远，变形程度就越大；为了对变形加以控制，测量中采用限制投影区域的办法，即将投影区域限制在中央子午线两侧一定的范围内，这就是所谓的分带投影。投影带一般分为 6° 带和 3° 带两种，如图 1-10 所示。

Although Gauss projection results in no angular deformation, it causes deformation in length and area. The further away from the central meridian, the greater the deformation. In order to control such deformations, the method to restrict the projection area is adopted in surveys, that is, to restrict the projection area within a certain range on both sides of the central meridian, which is the so-called zoning projection. There are generally the 6° zone and 3° zone, as shown in the Figure 1-10 below.

测图时，距离变形程度过大对测绘大比例尺地形图是不方便的。减小投影带边缘位置距离变形程度的方法之一就是缩小投影带的带宽。例如，可以选择采用 3° 带和 1.5° 带进行投影。其中，3° 带每带中央子午线经度与投影带 $N$ 的关系为

It is inconvenient in surveying and mapping for large-scale topographic maps with excessive distance deformation. One of the ways to reduce the distance deformation of the projection zone edge is to reduce the width of the projection zone. For example, projection may be given with 3° zones and 1.5° zones. The relationship between the central meridian longitude of each 3° zone and the projection zone $N$ is expressed as

$$L_0 = 3N \tag{1.2}$$

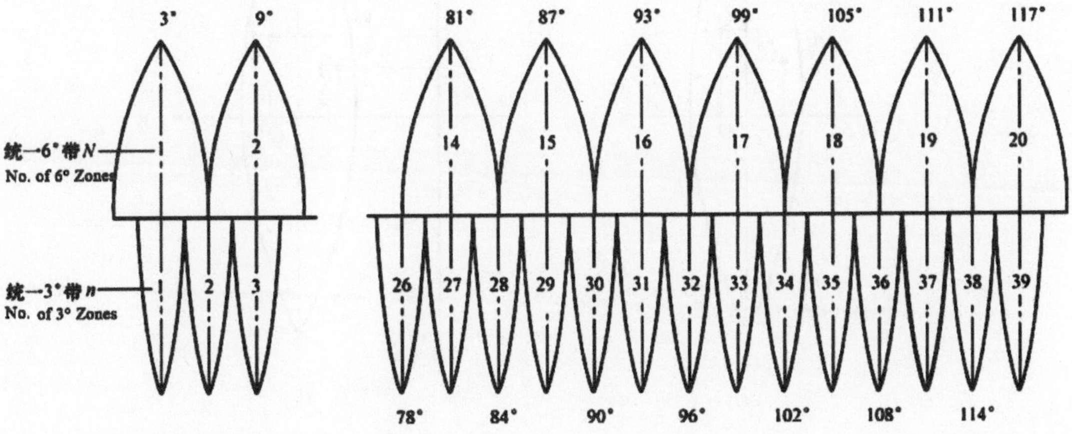


图1-10　6° 带和3° 带中央子午线

Figure 1-10 Central Meridian 6° Zones and 3° Zones

反之，已知地面任一点的经度 $L$，要求计算该点所在的 3° 带编号的公式为

Reversely, if the longitude $L$ of any point on the ground is known, the 3° zone number of this point is expressed in the formula

$$N = \text{Int}\ (L/3) \tag{1.3}$$

在我国领土范围内，6° 带与 3° 带的投影带号不重复。

The numbers of the 6° zones and the 3° zones projected are not overlapped within the territory of China.

通过高斯投影，将中央子午线的投影作为纵坐标轴，用 $x$ 表示，将赤道的投影作为横坐标轴，用 $y$ 表示，两轴的交点作为坐标原点，由此构成的平面直角坐标系称为高斯平面直角坐标系，如图 1-11 所示。每一个投影带，有一个对应的独立的高斯平面直角坐标系，区分各带坐标系则利用相应投影带的带号。

With the Gauss projection method, the central meridian projected is regarded as the axis of ordinates, expressed by $x$, the equator projected is regarded as the axis of abscissas, expressed by $y$, and the intersection point of the two axes is taken as the origin of coordinates. The plane rectangular coordinate system thus formed is called the Gauss plane rectangular coordinate system, as shown in the Figure 1-11. Each projection zone corresponds with an independent Gauss plane rectangular coordinate system, and these coordinate systems of all projection zones are distinguished by the corresponding zone number.

在每一投影带内，$y$ 坐标值有正有负，这在计算和使用中均不方便，为了使 $y$ 坐标都为正值，故将纵坐标轴向西平移 500 km(半个投影带的最大宽度不超过 500 km)，并在 $y$ 坐标前加上投影带的带号。如图 1-11 中 B 点位于 18 投影带，其自然坐标为 $x$=3 395 451 m，$y$=−82 261 m，它在 18 投影带中的高斯通用坐标则为 $X$=3 395 451 m，$Y$=18 417 739 m。

In each projection zone, there are positive

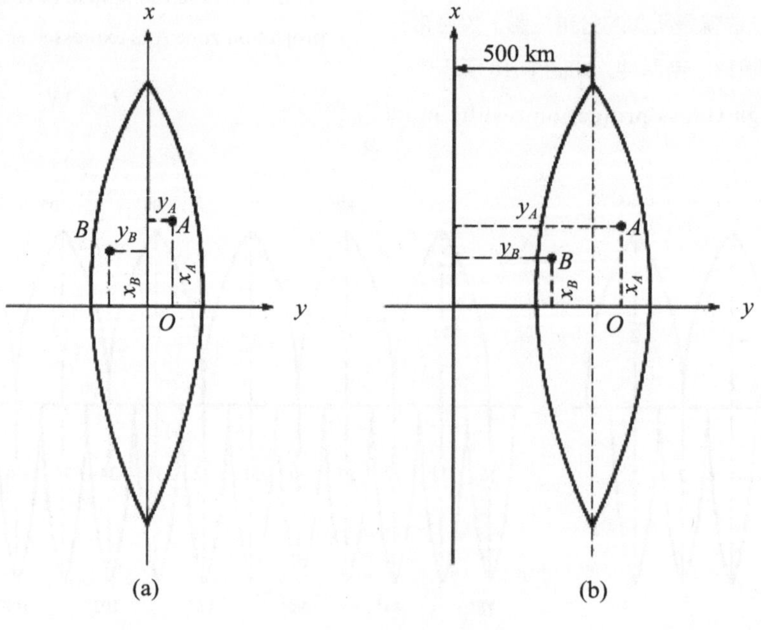


图1-11　高斯平面直角坐标系

Figure 1-11 Gauss Plane Rectangular Coordinate System

and negative $y$-coordinate values, which is inconvenient for computations and surveys. In order to make all the $y$-coordinate values positive, the axis of ordinates is translated westward for 500 km (the maximum width of half of the projection zone does not exceed 500 km), and the projection zone number is added to the $y$-coordinate values at the beginning. As shown in the Figure 1-11, point B is located in the No. 18 projection zone, with the natural coordinates of $x$=3 395 451 m and $y$=−82 261 m, and its Gauss general coordinates in the No. 18 projection zone are $X$=3 395 451 m and $Y$=18 417 739 m.

## 1.2.3 地面点高程位置的确定

## 1.2.3 Determination of Elevation of Ground Points

### （1）绝对高程

### (1) Absolute elevation

地面点到大地水准面的铅垂距离，称为该点的绝对高程，简称高程，或称海拔，用 $H$ 表示。地面点 $A$、$B$ 的绝对高程分别为 $H_A$、$H_B$。

The vertical distance from the ground point to the geoid is called the absolute elevation of the point, or the elevation or altitude for short, and is expressed by $H$. The absolute elevations of ground points A and B are $H_A$ and $H_B$ respectively.

目前，我国将根据青岛观象山验潮站1952—1979 年间的验潮资料确定的黄海平均海水面作为起算高程的基准面，称为“1985 国家高程基准”，该大地水准面为起算面，其高程为零。为了便于观测和使用，人们在青岛建立了我国的水准原点（国家高程控制网的起算点），其高程为 72.260 m，全国各地的高程都以其为基准进行测算。

At present, the mean sea level of the Yellow Sea defined according to the tide observation data at Qingdao Guanxiangshan Tidal Observatory from 1952 to 1979 is used as the datum plane for elevation, which is called “National Elevation Datum 1985”. This geoid is taken as the zero of elevation for surveying. In order to facilitate observation and surveys, the leveling origin of China (the zero point of the national vertical control network) was established in Qingdao, with an elevation of 72.260 m, as a benchmark of elevation determination across the country.

### （2）相对高程

### (2) Relative elevation

当测区附近尚无国家水准点，而引测绝对高程有困难时，可采用假定高程系统方法，即假定一个水准面作为高程基准面。这种由任意水准面起算的地面点高程即地面点至任意水准面的铅垂距离，称为相对高程，也称假定高程。如图 1-12 所示，$A$、$B$ 的相对高程为 $H'_A$、$H'_B$。有时为了使用方便，在某些工程测量中也常使用相对高程。

Under the circumstance that it is difficult to determine the absolute elevation by transfer due to the absence of a national benchmark near the survey area, an assumed elevation system may be adopted, that is, to assume a level surface as the datum plane for elevation. The elevation of a ground point measured from any level surface, i.e. the vertical distance from the ground point to any level surface, is called relative elevation, also known as assumed elevation. As shown in the Figure 1-12 below, the relative elevation of $A$ and $B$ is represented as $H'_A$ and $H'_B$. The relative elevation is also often used in engineering surveys for convenience.

### （3）高差

### (3) Elevation difference

地面两点之间的绝对高程或相对高程之差称为高差，用 $h_{AB}$ 表示，如图 1-13 所示。

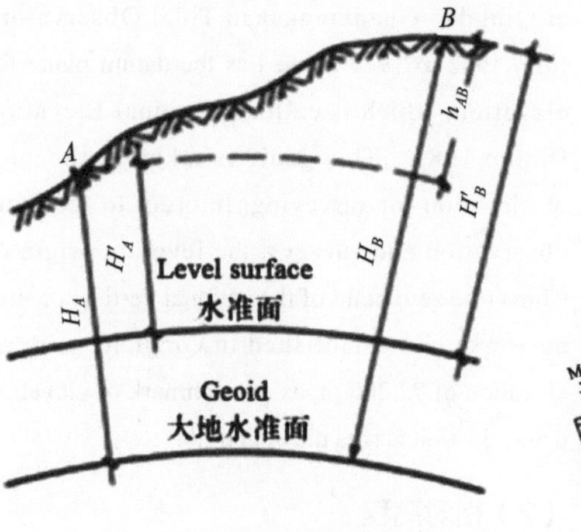


图1-12 地面点的高程

Figure 1-12 Elevation of Ground Point

The difference between two ground points at absolute elevation or relative elevation is called elevation difference, which is represented by $h_{AB}$, as shown in the Figure 1-13.

*A*、*B* 两点的高差为

The elevation difference between points *A* and *B* is expressed as

$$h_{AB}=H_B - H_A \tag{1.4}$$

从已知高程点对未知点进行高程测量时，都是先求出两点之间的高差，从而计算出未知点的高程。未知点比已知点高，其高差为正；反之，高差为负。

When measuring the elevation of a ground point based on another point with a known elevation, the elevation difference between the two points is measured first, and then the elevation is calculated for the surveyed points. If the surveyed point is located higher than the point with a known elevation, the elevation difference is expressed as a positive valve; otherwise, the elevation difference is negative.

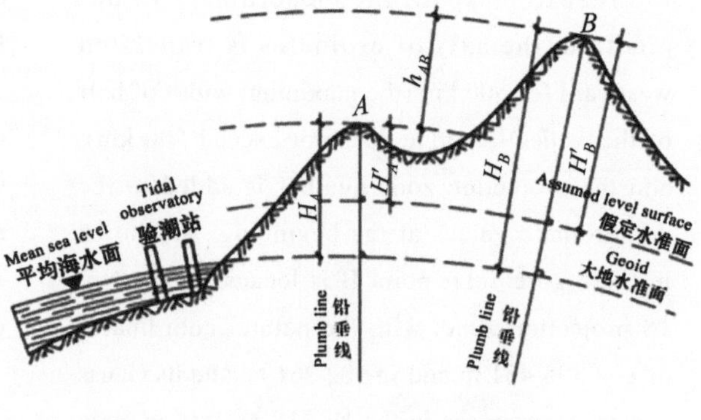


图1-13　高程与高差

Figure 1-13 Elevation and Elevation Difference

## 1.2.4 用水平面代替水准面的限度

## 1.2.4 Range of Replacing Level Surface with Horizontal Plane

当测区范围小，用水平面代替水准面所产生的误差不超过测量误差的容许范围时，可以用水平面代替水准面。但是在多大面积范围才容许这种代替，有必要加以讨论。为讨论方便，假定大地水准面为圆球面。

A level surface can be replaced with a horizontal plane in a small survey area when the error caused does not exceed the allowable range of observation errors. However, it is necessary to define the range to which such substitution is permitted. For the convenience of discussion , the geoid is assumed to be a spherical surface.

### （1）对距离的影响

### (1) Effect on distance

如图 1-14 所示，设地面上 *A*、*B*、*C* 三个点在大地水准面上的投影点是 *a*、*b*、*c*，用过 *a* 点的切平面代替大地水准面，则地面点水平面上的投影点是 $a$、$b'$、$c'$。设 $ab$ 的弧长为 $D$，$ab'$ 的长度为 $D'$，球面半径为 $R$，所对的圆心角则用水平长度 $D'$ 代替弧长。所产生的误差为

As shown in the Figure 1-14 below, assuming the projected points of points $A$, $B$ and $C$ on the ground are $a$, $b$ and $c$, the tangent plane passing through point $a$ is used to replace the geoid, and the projected points of the ground points on the horizontal plane are $a$, $b'$ and $c'$. Taking the arc length of ab as $D$, the length of $ab'$ as $D'$, and the spherical radius as $R$, the corresponding central angles are replaced by the horizontal length $D'$. The resulting error is expressed as

$$\Delta D=D'-D \tag{1.5}$$

将 $D=R\theta$、$D'=R\tan\theta$ 代入上式，整理后得

After substituting $D=R\theta$、$D'=R\tan\theta$ into the above formula, the following formula is derived as

$$\Delta D=R\left(\tan\theta-\theta\right) \tag{1.6}$$

将 $\tan\theta$ 展开为级数式

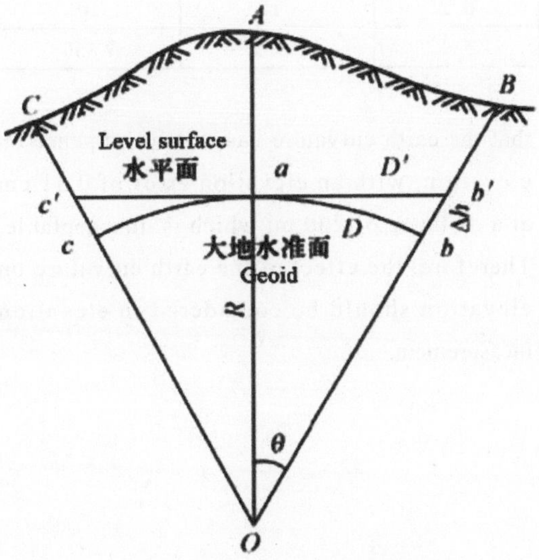


图1-14　水平面代替水准面的限度

Figure 1-14 Range of Replacing Level Surface with Horizontal Plane

$\tan\theta$ is developed as

$$\tan\theta = \theta + \frac{1}{3}\theta^2 + \frac{2}{15}\theta^3 + \cdots \tag{1.7}$$

因 $D$ 比 $R$ 小得多，$\theta$ 角很小，只取级数式前两项代入

Since $D$ is much smaller than $R$ and the angle $\theta$ is very small, only the first two series of the formula of series are substituted.

$$\Delta D=R\left(\theta + \frac{1}{3}\theta^2 - \theta\right) \tag{1.8}$$

将 $\theta=\frac{D}{R}$ 代入上式得

After substituting into the above formula, the following formula is derived as

$$\frac{\Delta D}{D} = \frac{D^2}{3R^2} \tag{1.9}$$

取 $R$=6 371 km，用不同的 $D$ 值代入式（1.9）得到表 1-1 所示的结果。从表 1-1 中可知，当两点相距 10 km 时，用水平面代替大地水准面产生的长度误差为 0.8 cm，相对误差为 1/1 220 000，相当于精密测距精度的 1/1 000 000。所以，在半径为 10 km 范围的测区进行距离测量时，可以用水平面代替大地水准面。

Taking $R$=6 371 km and substituting different $D$ values into Formula (1.9), the results shown in Table 1-1 are generated. It can be seen from Table 1-1 that when the two points are 10 km apart, the length error caused by replacing the geoid with a horizontal plane is 0.8 cm, and the relative error is 1/1 220 000, which is equivalent to 1/1 000 000 of the accuracy for precise distance measurement. Therefore, the geoid can be replaced by a horizontal plane for the distance measurement in the survey area with a radius of 10 km.

表 1-1　水平面代替大地水准面对距离的影响

Table 1-1 Effect of Replacing Level Surface with Horizontal Plane on Distance

| $D$/km | $\Delta D$/cm | $\Delta D/R$ |
|---|---|---|
| 5 | 0. 1 | 1/4 870 000 |
| 10 | 0. 8 | 1/1 220 000 |
| 20 | 6. 6 | 1/304 000 |
| 50 | 102. 7 | 1/48 700 |

## （2）对高程的影响

### (2) Effect on Elevation

在图 1-14 中，以大地水准面为基准的 $B$ 点绝对高程 $H_B = Bb$，用水平面代替大地水准面时，$B$ 点的高程 $H'_B = Bb'$，两者之差就是对高程的影响，也称为地球曲率的影响。由图可知

In Figure 1-16, the absolute elevation of point $B$ based on the geoid is $H_B = Bb$, and the elevation of point $B$ is $H'_B = Bb'$, when replacing the level surface with a horizontal plane, resulting in a difference between the two results, which is regarded as the effect on the elevation, also known as the influence of the earth curvature. It can be seen from the figure that

$$\Delta h = Bb - Bb' = Ob' - Ob = R\sec\theta - R = R(\sec\theta - 1)$$

对于不同的 $D$ 值，产生的高程影响见表 1-2。

For different $D$ values, the resulting elevation effect is shown in Table 1-2.

表 1-2　水平面代替大地水准面对高程的影响

Table 1-2 Effect of Replacing Level Surface with Horizontal Plane on Elevation

| $D$/km | 0. 05 | 0. 1 | 0. 2 | 1 | 10 |
|---|---|---|---|---|---|
| $\Delta h$/mm | 0. 2 | 0. 8 | 3. 1 | 78. 5 | 7 850 |

表 1-2 的计算结果表明，地球曲率对高程的影响较大，距离 200 m 就有 0.31 cm 的高程误差，这是不允许的。因此，进行高程测量时，应考虑地球曲率对高程的影响。

The calculation results in Table 1-2 show that the earth curvature has a great influence on elevation, with an elevation error of 0.31 cm at a distance of 200 m, which is unacceptable. Therefore, the effect of the earth curvature on elevation should be considered in elevation measurement.

# 1.3 测量的基本工作

# 1.3 Basic Work of Surveying

## 1.3.1 测量的基本内容

## 1.3.1 Basic Content of Surveying

测量工作的主要目的是确定点的坐标和高程。实际测量工作常常不是直接测量点的坐标和高程，而是观测坐标和高程已知的点，与坐标和高程未知的待定点之间的几何位置关系，然后推算出待定点的坐标和高程。

The main purpose of surveying is to determine the coordinates and elevation of points. The actual survey practice is to observe the geometric position relationship between the point with known coordinates and elevation and the surveyed point with unknown coordinates and elevation, and then calculate the coordinates and elevation of the surveyed point, instead of direct measurement of coordinates and elevation.

如图 1-15 所示，设 *A*、*B* 为坐标和高程已知的点，*C* 为待定点，三点在投影水平面上的投影位置分别是 *a*、*b*、*c*，在△ *abc* 中，*ab* 边的长度是已知的，只要测量出一条未知边的边长和一个水平角（或两个水平角，或两个未知边边长），就可以推算出 *C* 点的坐标。可见，测定地面点的坐标主要是测量水平距离和水平角。

As shown in the Figure 1-15, *A* and *B* are the points with known coordinates and elevations and *C* is the surveyed point, the projected positions of the three points on the projection horizontal plane are *a*, *b* and *c* respectively, and the length of the side *ab* is known in △ *abc*. As long as the length of an unknown side and a horizontal angle (or two horizontal angles or two unknown side lengths) are measured, the coordinates of point *C* can be calculated. It can be seen that the coordinates of a ground point are determined mainly by measuring the horizontal distance and horizontal angle.

欲求 *C* 点的高程，则要测量出高差（或两个水平角，或两个未知边边长），然后推算出 *C* 点的高程，所以，测定地面点高程主要是测量高差。

To get the elevation of point *C*, the elevation difference(or two horizontal angles or two unknown side lengths) needs to be measured, and then the elevation of point *C* can be calculated. Therefore, the elevation of a ground point is determined mainly by measuring the elevation difference.

因此，高差测量、角度测量、距离测量是测量工作的基本内容。

To sum up, the measurement of elevation difference, angle, and distance constitutes the main part of survey practice.

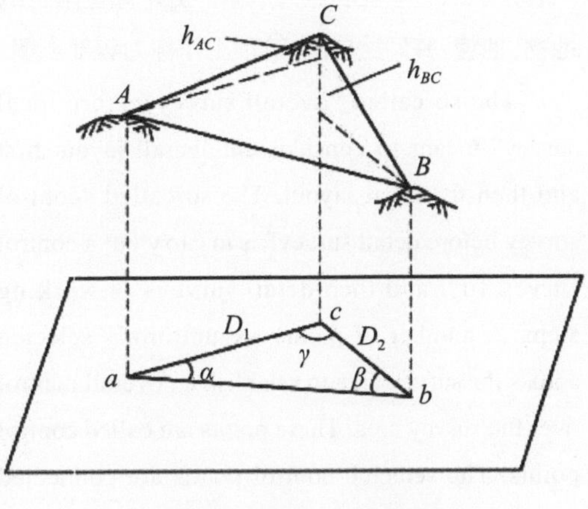


图1-15　角度、距离测量

Figure 1-15 Angle and Distance Measurement

## 1.3.2 测量的基本原则

## 1.3.2 Basic Principle of Surveying

为了克服误差的传播和累积对测量成果造成的影响，测量工作必须遵循一定的原则进行。这一原则就是“先整体后局部，先控制后碎部，高精度控制低精度”，即先进行整体的、精度较高的控制测量，再进行局部的、精度较低的碎部测量。

In order to overcome the influence of error propagation and accumulation on the survey results, a certain basic principle must be followed in actual surveys, i.e., “overall survey before local survey, control survey before detail survey, survey at a high precision prevailing over survey at a low precision”, which means to carry out the overall control survey at a high precision first, and then carry out the local detail survey at a low precision.

所谓“先整体后局部”就是在布局上先考虑整体，再考虑局部。所谓“先控制后碎部”就是在工作步骤上先进行控制测量，再进行碎部测量。从整体出发，先在整个测区范围内均匀选定若干数量的点，以控制整个测区，这些点称为控制点。选定的控制点按照一定的方式联结成网形，称为控制网。以较精密的方法测定网中各个控制点的平面位置和高程，这项工作称为控制测量。然后分别以这些控制点为依据，测定点位附近的地物、地貌，并勾绘成图，这项工作称为碎部测量。

The so-called “overall survey before local survey” means to consider the overall layout first and then the local layout. The so-called “control survey before detail survey” is to carry out a control survey first and then detail surveys in working steps. A number of points are uniformly selected across the survey area to establish an overall control over the survey area. These points are called control points. The selected control points are connected into a network in a certain way, which is called the control network. The plane position and elevation of each control point in the network are measured by a precise method, which is called a control survey. Then the ground objects and land features near the surveyed points are measured and mapped based on these control points, which is called a detail survey.

按照“先整体后局部，先控制后碎部”的原则实施测图时，由于建立了统一的控制系统，各个控制点的坐标和高程是通过平差处理而得到的，各个控制点乃至以各个控制点为测站所做的碎部测量都具有相同的精度，从而有效地防止了误差累积。同时，碎部测量又是在各个控制点上独立进行的，这将大大提高碎部测量的机动性和灵活性，尤其对大面积测区的分幅测图来说，不仅为分幅测图作业提供了便利，也有效地保证了各相邻图幅的拼接和使用。

Mapping is carried out according to the principle of “overall survey before local survey, control survey before detail survey”. Since a unified control system is established, the coordinates and elevation of the control points are generated by adjustment processing, the detail surveys of the control points and even the surveys made with the control points as datums are made with the same accuracy, thus effectively preventing the accumulation of errors. Moreover, the detail surveys are carried out independently on the control points, to greatly improve the maneuverability and flexibility of the detail surveys, which particularly is not only convenient for divided mapping of large survey areas, but also effectively ensures the splicing of adjacent map sheets.

“先整体后局部，先控制后碎部”的原则同样适用于施工测量。为了将图上设计的建筑物、构筑物放样到实地去，同样应从整体出发，首先建立施工控制网，然后根据控制点和放样数据来测设建筑物、构筑物的细部点。

The principle of “overall survey before local survey, control survey before detail survey” is also applicable to construction surveys. In order to set out the buildings and structures designed on the drawing to the site, a construction control network needs to be established first from an overall perspective, and then the detail points of the buildings and structures are set out according to the control points and setting-out data.

应当指出，测量工作有“外业”和“内业”之分，利用测量仪器和工具在现场所进行测角、测高、测距等测量工作称为测量外业；对观测数据、资料在室内进行计算、整理和绘图等工作称为测量内业。外业和内业共同决定着测量结果的质量，工作环节上的任何一处失误，都将给后续的一系列工作造成严重影响。因此无论外业或内业工作，都必须坚持“边工作边检核”“步步工作有检核”的工作原则。同时，测量工作又是一项复杂的集体劳动，任何疏忽和麻痹大意都可能导致不合格结果出现，造成部分或整体返工，所以，要求测量人员具有团结协作的工作作风以及严谨细致的工作态度是十分重要和必要的。

It should be noted that surveys are divided into “field work” and “office work”. The former refers to the surveys carried out on the spot with measuring instruments and tools, such as angle measurement, elevation measurement, and distance measurement. The latter includes calculation, sorting, and mapping based on observation data and information in the office. The field work and office work jointly contribute to the quality of the survey results, and any mistake in any work steps would bring a serious impact on subsequent work. Therefore, the working principle of “checking while working” and “checking in each working step” must be adhered to in both the field work and office work. In addition, in the complex surveying process of collective efforts, any negligence and carelessness may lead to non-conforming results, resulting in partial or overall rework. Therefore, it is very important and necessary for the surveyors to keep a work style of solidarity and cooperation and a rigorous and careful work attitude.

# 本章小结

## Chapter Summary

本章主要讲述了建筑工程测量的任务、地面点位的确定、测量工作的基本要求和原则、测量误差及测量精度指标等内容。

This chapter mainly introduces the tasks of architectural engineering surveying, determination of ground points, basic requirements and principles of surveying, observation errors and observation accuracy indicators, etc.

1. 建筑工程测量的主要任务是大比例尺地形图的测绘、施工放样和竣工测量以及建（构）筑物的变形观测。

1. The main tasks of architectural engineering surveying include surveying and mapping for large-scale topographic maps, construction setting-out and acceptance survey, and deformation observation of buildings (structures).

2. 在测量工作中，可用地理坐标和平面直角坐标表示地面点位置的坐标。为了使用方便，常采用平面直角坐标来表示地面点位。地面点到大地水准面的铅垂距离称为该点的绝对高程。由任意水准面起算的地面点高程，即地面点到任意水准面的铅垂距离，称为相对高程。地面两点之间的绝对高程或相对高程之差称为高差。

2. In actual surveys, ground points can be represented by geographical coordinates and plane rectangular coordinates, and the latter is more frequently used for the sake of convenience. The vertical distance from a ground point to the geoid is called the absolute elevation of the point. The elevation of a ground point measured from any level surface, i.e., the vertical distance from the ground point to any level surface, is called the relative elevation of the point. The difference between two ground points at absolute elevation or relative elevation is called elevation difference.

3. 测量工作必须遵循"先整体后局部，先控制后碎部"和"边工作边检核"的原则。

3. The principles of "overall survey before local survey, control survey before detail survey" and "checking while working" must be abided by in actual surveys.

4. 常用的衡量测量精度的指标有中误差、极限误差和相对误差。

4. The common indicators to measure observation accuracy include mean square error, limit error and relative error.

# 复习思考题

# Review Questions

## 一、填空题

## I. Fill in the Blanks

1. 测量学的基本任务包括______________和______________两个方面。

1. Basic tasks of surveying include two aspects, ______________and______________.

2. 在测量工作中，可用______________和______________表示地面点位置的坐标。

2. In actual surveys, ground points can be represented by______________and______________.

3. 在高斯平面直角坐标系中，中央子午线的投影为______________坐标轴。

3. In the Gauss plane rectangular coordinate system, the central meridian projected is regarded as the axis of ______________.

4. 平面直角坐标系中，象限按______________方向编号，*x* 轴与 *y* 轴互换。

4. In the plane rectangular coordinate system, quadrants are numbered in the direction of______________, and *x*-axis and *y*-axis are interchangeable.

5. 我国以______________平均海水面作为起算高程的基准面。

5. The mean sea level of______________is used as the datum plane for elevation in China.

6.______________、______________、______________是测量工作的基本内容。

6.______________,______________,______________constitutes the basic content of surveying.

## 二、选择题

## II. Choice Questions

1. 我国现在采用的水准原点设在（ ）。

1.The leveling origin currently adopted in China is set at ( ).

A. 北京 B. 上海 C. 青岛 D. 西安

A. Beijing B. Shanghai C. Qingdao D. Xi'an

2. 在高斯平面直角坐标系中，纵轴为（ ）。

2.In the Gauss plane rectangular coordinate system, the ordinate axis is ( ).

A. *X* 轴，向东为正 B. *Y* 轴，向东为正

A. *X*-axis, positive eastward B. *Y*-axis, positive eastward

C. *X* 轴，向北为正 D.*Y* 轴，向北为正

C. *X*-axis, positive northward D. *Y*-axis, positive northward

3. 某点所在的 6° 带的高斯坐标值为 *X*=366 712. 48 m，*Y*=21 331 229.75 m，则该点位于（ ）。

3.If the Gauss coordinate values of the 6° zone where a point is located are *X*=366 712.48 m and *Y*=21 331 229.75 m, this point is located at ( ).

A. 21 带，在中央子午线以东 B. 36 带，在中央子午线以东

A. zone 21, east of the central meridian B. zone 36, east of the central meridian

C. 21 带，在中央子午线以西 D. 36 带，在中央子午线以西

C. zone 21, west of the central meridian D. zone 36, west of the central meridian

4. 当测区附近尚无国家水准点，而引测绝对高程有困难时，可采用假定高程系统，即假定一个

水准面作为（　）。

4.Under the circumstance that it is difficult to determine the absolute elevation by transfer due to the absence of a national benchmark near the survey area, an assumed elevation system may be adopted, that is, to assume a level surface as (　).

A. 绝对高程　　B. 相对高程　　C. 高差　　D. 标高

A. absolute elevation　　B. relative elevation　　C. elevation difference　　D. elevation

## 三、简答题

### III. Short Answer Questions

1. 建筑工程测量的任务是什么？

1.What are the tasks of architectural engineering surveying?

2. 用水平面代替大地水准面，对水平距离和高程分别有何影响？

2.What are the effects of replacing a level surface with a horizontal plane on horizontal distance and elevation respectively?

3. 什么是绝对高程？什么是相对高程？什么是高差？

3.What is absolute elevation? What is relative elevation? What is elevation difference?

4. 测量工作的基本原则是什么？

4.What is the basic principle of surveying?

5. 误差的产生原因有哪些？

5.What are the causes of errors?

6. 系统误差和偶然误差有什么不同？在测量工作中应如何处理这两种误差？

6.What is the difference between systematic and accidental errors? How to deal with these two kinds of errors in surveying?

7. 什么是测量精度？衡量测量精度的指标有哪些？

7.What is observation accuracy? What are the indicators for measuring observation accuracy?

2

# 水准测量

# Leveling

**学习目标：**

**Learning Objectives:**

通过本章的学习，了解水准测量的原理、仪器和工具；熟悉水准仪的构造及各部件的名称、作用以及检验与校正方法；掌握水准仪的基本操作方法，水准线路测量的外业、内业工作方法。

Through the study of this chapiter, to learn about the principles, instruments and tools of leveling; to get familiar with the structure of levels, the name and function of their components, and the check and calibration methods of levels; to master the basic operation methods of levels and the methods of field work and office work in route leveling.

**能力目标：**

**Ability Objectives：**

具备灵活进行水准测量的能力，能够熟练操作水准仪，并能正确进行水准仪的检验与校正，能够运用水准测量方法进行地面点高程的测量与测设。

Be able to flexibly conduct leveling, skillfully operate, as well as correctly check and calibrate levels, and apply the leveling methods to measure the elevation of ground points and carry out setting-out.

# 2.1 水准测量的原理

## 2.1 Principles of Leveling

测量地面上各点高程的工作称为高程测量。根据所使用的仪器和施测方法以及精度要求的不同，高程测量可分为水准测量、三角高程测量、气压高程测量和 GPS 测量等。水准测量是高程测量中最基本的精密度较高的一种测量方法，被广泛应用于国家高程控制测量、工程勘测和施工测量工作中。

The work of measuring the elevation of ground points is called elevation measurement. Elevation measurement is classified into leveling, trigonometric leveling, barometric leveling and GPS leveling, according to different instruments, surveying methods and accuracy requirements. Among them, leveling is the most basic surveying method with high precision, which is widely used in national vertical control surveys, engineering surveys and construction surveys.

水准测量指利用水准仪提供的水平视线，根据水准仪在两点竖立的水准尺上的读数，先求得两点之间的高差，然后根据已知点的高程推算出未知点的高程。

Taking advantage of the horizontal sight provided by a level, leveling is to calculate the elevation difference between two points according to the readings on the leveling rods erected at the two points, and then calculate the elevation of the surveyed point according to the point with a known elevation.

如图 2-1 所示，已知 $A$ 点高程 $H_A$，测定 $B$ 点的高程 $H_B$。施测时，在 $A$、$B$ 两点之间约等距离处安置水准仪，并在 $A$、$B$ 两点分别垂直竖立水准尺（也称水准标尺），先照准 $A$ 点水准尺，根据水准仪提供的水平视线在 $A$ 点水准尺上的读数 $a$，再照准 $B$ 点的水准尺，保持同一水平视线，读出读数 $b$。

The figure 2-1 shows the measurement of the elevation $H_B$ for point $B$, according to the known elevation $H_A$ of point $A$. Before measurement, levels are placed at an approximately equal distance between points $A$ and $B$, and leveling rods (also called leveling staffs) are erected vertically at points $A$ and $B$ respectively. First, the leveling rod at point $A$ is aimed at, and then the leveling rod at point $B$ is aimed at according to the reading $a$ of the leveling rod at point $A$ acquired along the horizontal sight provided by the levels, and the reading $b$ is acquired along the same horizontal sight.

如图 2-1 中箭头所示，测量的前进方向由已知高程的 $A$ 点向待测高程的 $B$ 点。$A$ 点称为后视点，所立水准尺为后视尺，尺上读数 $a$ 称为后视读数；$B$ 点称为前视点，所立水准尺为前视尺，尺上读数 $b$ 为前视读数。

As shown by the arrow in Figure 2-1, the forward direction in the survey is from point $A$ with the known elevation to surveyed point $B$. Point $A$ is called the backsight point, the leveling rod erected is the backsight leveling rod, and the reading $a$ on the leveling rod is called the backsight reading; point $B$ is called the foresight point, the leveling rod erected is the foresight leveling rod, and the reading $b$ on the leveling rod is the foresight reading.

地面上已知 $A$ 点与待测 $B$ 点两点之间高差始终等于后视读数 $a$ 减去前视读数 $b$。

The elevation difference between the known point $A$ and the surveyed point $B$ on the ground is always equal to the backsight reading $a$ minus the foresight reading $b$.

$h_{AB}$= 后视读数 − 前视读数 = $a-b$ （2.1）

$h_{AB}$= backsight reading − foresight reading = $a-b$ （2.1）

在实际工作中，通常 $A$、$B$ 两点相距较远或高差较大，仅安置一次仪器难以测得两点的高差，此时需连续设站进行观测。如图 2-2 所示，在 $A$、$B$ 两点之间增设若干个临时立尺点，将 $AB$ 划分为 $n$ 段，逐段安置水准仪进行水准测量。

In actual surveys, points $A$ and $B$ are usually far apart or greatly varied in elevation, and it is difficult to measure the elevation difference between the two points with just one observation station, so it is necessary to set up continuous observation stations. As shown in Figure 2-2, a number of temporary points of leveling rods are added between points $A$ and $B$ to divide the distance between points $A$ and $B$ into $n$ sections, and levels are erected section by section for leveling.

将安置仪器的位置称为测站，在每一测站上进行水准测量，得到各测站的后视读数和前视读数分别为 $a_1$、$b_1$、$a_2$、$b_2$、$a_n$、$b_n$，则各测站测得的高差为

The position where the measuring instrument is placed is called an observation station, and leveling is carried out on each observation station to obtain the backsight readings and foresight readings of all the observation stations, which are $a_1$, $b_1$, $a_2$, $b_2$, $a_n$ and $b_n$ respectively, so the elevation difference measured by each observation station is as follows

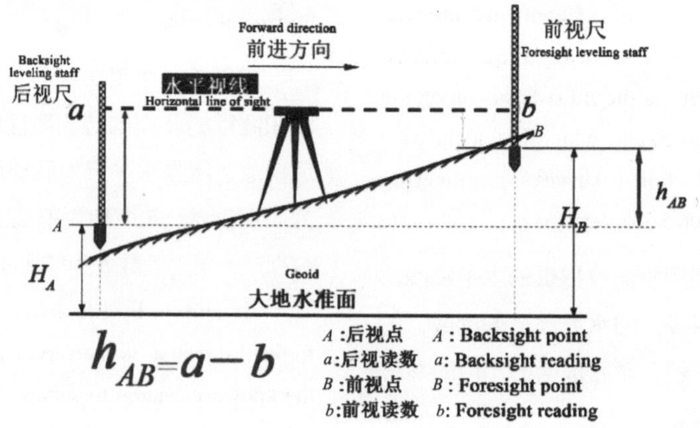


图2-1 水准测量

Figure 2-1 Leveling

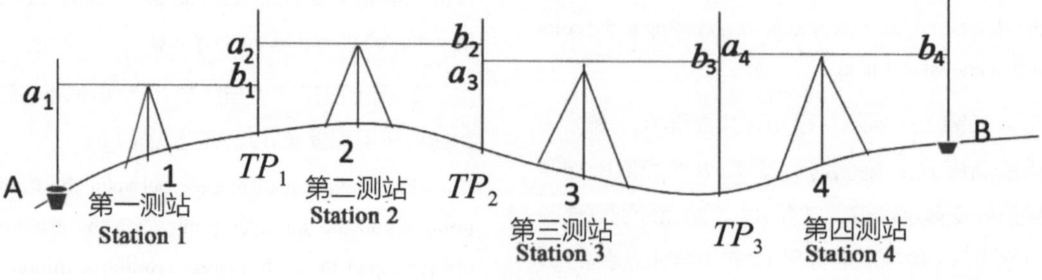


第一测站

图2-2 连续水准测量

Figure 2-2 Continuous Leveling

第一测站　$h_1=a_1-b_1$

Station 1　$h_1=a_1-b_1$

第二测站　$h_2=a_2-b_2$

Station 2　$h_2=a_2-b_2$

…　…

第 $n$ 测站　$h_n=a_n-b_n$

Station $n$　$h_n=a_n-b_n$

$A$、$B$ 两点的高差为

Elevation difference between points $A$ and $B$ can be expressed as

$$h_{AB}=h_1+h_2+\ldots+h_n \tag{2.2}$$

若 $A$ 点高程已知，则 $B$ 点的高程为

If the elevation of point $A$ is known, the elevation of point $B$ is expressed as

$$H_B=H_A+h_{AB} \tag{2.3}$$

在水准测量中，$A$、$B$ 两点之间的临时立尺点仅起传递高程的作用，这些点称为转点，通常以 $TP$ 表示，如图 2-2 中的 $TP_1$、$TP_2$、$TP_3$。

The temporary points of leveling rods between points $A$ and $B$ only play a role of elevation transfer, and these points are called turning points usually expressed by $TP$, shown as $TP_1$, $TP_2$, $TP_3$ in the Figure 2-2.

# 2.2　水准测量的仪器和工具

# 2.2 Instruments and Tools of Leveling

## 2.2.1 水准仪简介

## 2.2.1 Introduction to Levels

水准仪是进行水准测量的主要仪器，它可以提供水准测量所必需的水平视线。目前通用的水准仪从构造上可分为两大类：一类是利用水准管来获得水平视线的水准管水准仪，其主要形式为“微倾式水准仪”；另一类是利用补偿器来获得水平视线的自动安平水准仪，如图 2-3。另外，还有一种新型水准仪——电子水准仪，它配合条纹编码尺，利用数字化图像处理的方法，可自动显示高程和距离，使水准测量实现自动化。

As the main instrument for leveling, levels are used to provide the necessary horizontal line of sight for leveling. The currently common levels are divided into two categories in terms of structure: one is to use a level tube to generate a horizontal line of sight, mainly represented by the “tilt level”; the other is the “automatic compensated level” which generates a horizontal line of sight by a compensator (Figure 2-3) . In addition, there is a new type of level, the electronic level, which can automatically display elevation and distance by leveraging the digital image processing method and in combination with a stripe coding rod, to enable automatic leveling.

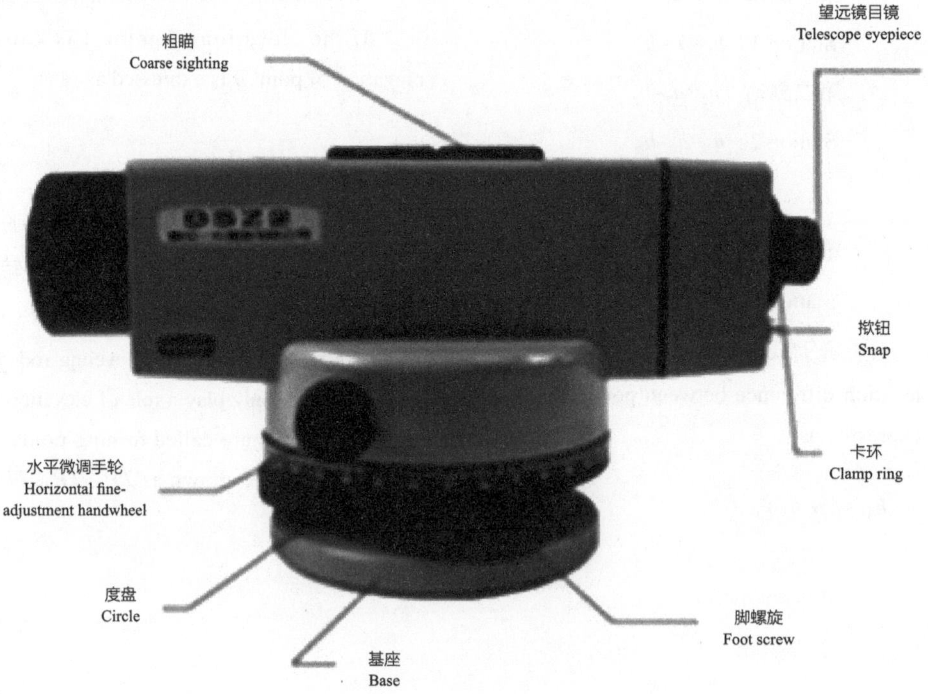


图2-3　自动安平水准仪

Figure 2-3 Automatic Compensated Level

我国的水准仪系列标准分为 $DS_{05}$、$DS_1$、$DS_3$ 和 $DS_{20}$ 四个等级。D是大地测量仪器的代号，S是水准仪的代号，均取“大”和“水”两个字的汉语拼音的首字母。角码的数字表示仪器的精度，即每千米往、返水准测量误差，以mm为单位。$DS_{05}$、$DS_1$、$DS_3$ 和 $DS_{20}$ 水准仪每千米往、返测量高差中数的中误差分别为 ±0.5 mm、±1 mm、±3 mm、±20 mm。

The domestic levels are divided into four series, i.e. $DS_{05}$, $DS_1$, $DS_3$ and $DS_{20}$ to distinguish the accuracy levels. D represents a geodetic instrument and S represents a level, both derived from the first letter of the Chinese pinyin of “Da” and “Shui”. The number of the angle code indicates the instrument accuracy in mm, that is, the error per kilometer of double-run leveling. The mean square error of the elevation difference of $DS_{05}$, $DS_1$, $DS_3$ and $DS_{05}$ levels is ±0.5 mm, ±1 mm, ±3 mm and ±20 mm respectively per kilometer of double-run leveling.

其中 $DS_{05}$ 和 $DS_1$ 为精密水准仪，用于精密水准测量，主要用于国家一、二等水准测量和精密工程测量；$DS_3$ 用于一般水准测量，主要用于国家三、四等水准测量；$DS_{20}$ 则用于简易水准测量，主要用于常规工程建设测量。在工程建设中，使用最多的普通水准仪是 $DS_3$ 水准仪。

Among them, $DS_{05}$ and $DS_1$ are precision levels, which are mainly used for national first-order and second-order leveling and precision engineering surveys; $DS_3$ is used for general leveling, which is mainly used for national third-order and fourth-order leveling; $DS_{20}$ is used for simple leveling, which is mainly used for conventional engineering construction surveys. $DS_3$ levels are most commonly used in engineering construction.

## 2.2.2 $DS_3$ 型微倾式水准仪的构造

### 2.2.2 Structure of $DS_3$ Tilt Level

图 2-4 所示为在一般水准测量中使用较多的 $DS_3$ 型微倾式水准仪，它由三个主要部分组成：望远镜，可以提供视线，并可读出远处水准尺上的读数；水准器，用于指示仪器或视线是否处于水平位置；基座，用于置平仪器，支承仪器的上部并能使仪器的上部在水平方向转动。

Figure 2-4 shows the $DS_3$ tilt level, which is widely used in general leveling and consists of three main parts: a telescope to generate a line of sight and acquire readings on a remote leveling rod; a bubble to indicate whether the instrument or line of sight is in a horizontal position; a base used to place the instrument horizontally, support the upper part of the instrument and keep the upper part of the instrument in the horizontal rotation.

水准仪各部分的名称如图 2-4 所示。基座上有三个脚螺旋，调节脚螺旋可使圆水准器的气泡移至中央，使仪器粗略整平。望远镜和管水准器与仪器的竖轴联结成一体，竖轴插入基座的轴套内，可使望远镜和管水准器在基座上绕竖轴旋转。制动螺旋和微动螺旋用来控制望远镜在水平方向的转动。制动螺旋松开时，望远镜能自由旋转；旋紧时，望远镜则固定不动。旋转微动螺旋可使望远镜在水平方向作缓慢的转动，但只有在制动螺旋旋紧时，微动螺旋才能起作用。

The name of the components of a level is shown in Figure 2-4. The bubble of a circular level can be moved to the center by adjusting the three foot screws on the base, so as to roughly level the instrument. The telescope and bubble tube are integrated with the vertical axis of the instrument, and the vertical axis is inserted into the shaft sleeve of the base, so that the telescope and bubble tube can be rotated around the vertical axis on the base. The clamp and fine-motion screws are used to control the horizontal rotation of the telescope. The clamp screw is unfastened to enable free rotation of the telescope, fastened to fix the telescope. The fine-motion screw is rotated to enable slow horizontal rotation of the telescope, which cannot work until the clamp screw is fastened.

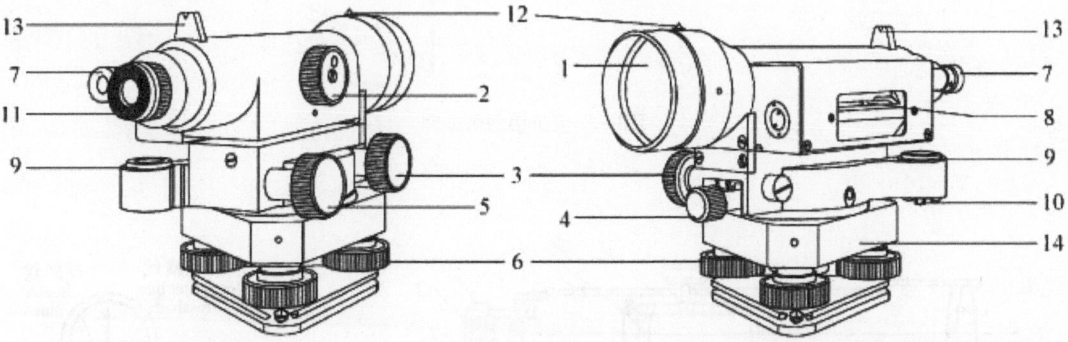


1—物镜；2—物镜调焦螺旋；3—微动螺旋；4—制动螺旋；5—微倾螺旋；6—脚螺旋；7—气泡观察镜；8—管水准器；9—圆水准器；10—校正螺丝基座；11—目镜；12—准星；13—照门；14—基座。

1-objective lens; 2-focusing screw of objective lens; 3-fine-motion screw; 4-clamp screw; 5-tilt screw; 6-foot screw; 7-bubble sight glass; 8-bubble tube; 9-circular bubble; 10-correction screw base; 11-eyepiece; 12-front sight; 13-rear sight; 14-base.

图2-4　$DS_3$型微倾式水准仪

Figure 2-4 $DS_3$ Tilt Level

旋转微倾螺旋可使望远镜连同管水准器作俯仰微量的倾斜，从而可使视线精确整平。因此，这种水准仪称为微倾式水准仪。

The tilt screw is rotated to tilt the telescope and the bubble tube slightly, until the line of sight is accurately leveled, from which the name of this kind of level is derived.

下面说明微倾式水准仪主要的部件——望远镜、水准器的构造和性能。

The construction and performance of the main components of a tilt level are described below, i.e. telescope and bubble.

## （1）望远镜

### (1) Telescope

最简单的望远镜是由物镜和目镜组成的。物镜的作用是使物体在物镜的另一侧构成一个倒立的实像；目镜的作用是使这一实像在同一侧形成一个放大的虚像（图 2-5）。为了使物像清晰并消除单透镜的一些缺陷，物镜和目镜都是用两种不同材料的透镜组合而成的（图 2-6）。

The simplest telescope consists of an objective lens and eyepiece. The objective lens is used to cast an inverted real image on the other side of the objective lens; the eyepiece is used to form an enlarged virtual image on the same side (Figure 2-5). In order to make the object image clear and eliminate some defects of a single lens, both the objective lens and eyepiece are made of two different materials (Figure 2-6).

测量仪器上的望远镜还必须有一个十字丝分划板，它是刻在玻璃片上的一组十字丝，被安装在望远镜筒内靠近目镜的一端。水准仪上

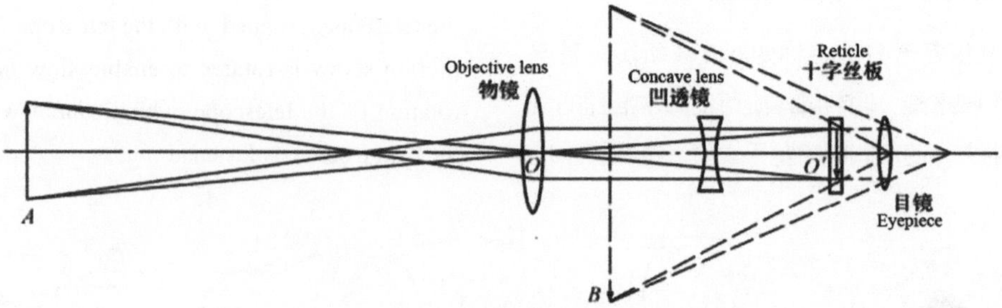


图2-5 望远镜成像原理

Figure 2-5 Telescope Imaging Principle

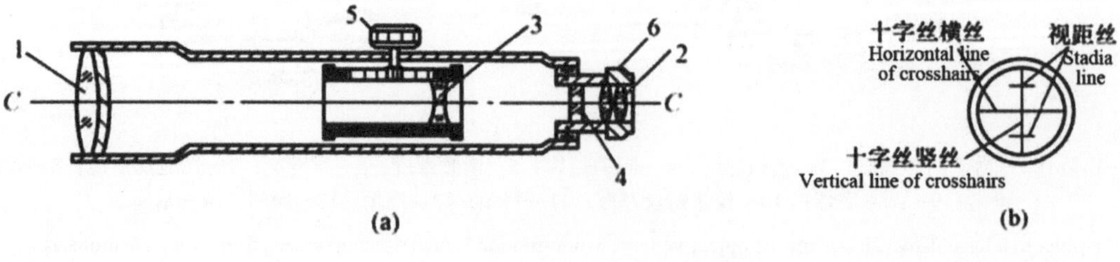


1—物镜；2—目镜；3—调焦透镜；4—十字丝分划板；5—调焦螺旋；6—目镜调焦螺旋。

1-objective lens; 2-eyepiece; 3-focusing lens; 4-crosshair reticle; 5-focusing screw; 6-eyepiece focusing screw.

图2-6 望远镜构造

Figure 2-6 Telescope Structure

十字丝的图形如图 2-7 所示，水准测量中用它中间的横丝或楔形丝读取水准尺上的读数。十字丝交点和物镜光心的连线称为视准轴，也就是视线。视准轴是水准仪的主要轴线之一。

The telescope on measuring instruments must also be provided with a crosshair reticle, which is a set of crosshairs engraved on the glass sheet and installed near the eyepiece end in the telescope barrel. Crosshairs on levels are shown in Figure 2-7. Their horizontal line or wedged line is used to take readings on the leveling rod. The line connecting the intersection point of crosshairs and the optical center of the objective lens is called the collimation axis, that is, the line of sight. The collimation axis is one of the main axes of levels.

图2-7 十字丝

Figure 2-7 Crosshairs

为了能准确地照准目标或读数，望远镜内必须能同时看到清晰的物像和十字丝。为此必须使物像落在十字丝分划板平面上。为了使离仪器不同距离的目标能成像于十字丝分划板平面上，望远镜内还必须安装一个调焦透镜（图 2-6）。当观测不同距离处的目标时，可旋转调焦螺旋改变调焦透镜的位置，从而观测者能在望远镜内清晰地看到十字丝和所要观测的目标。

In order to accurately aim at targets or take readings, the telescope must present a clear sight of both the object image and crosshairs. For this purpose, the object image must be cast on the reticle plane. A focusing lens must also be installed in the telescope so that targets distanced differently from the level can be imaged on the reticle plane (Figure 2-6). The focusing screw is rotated to change the position of the focusing lens, adapting to different distances of targets to be observed, so that the crosshairs and targets can be clearly seen in the telescope.

从以上内容可以看出，望远镜的各项性能是相互制约的。例如增大放大率可增强分辨率，可提高观测精度，但减小了视场角度和降低了亮度，就不利于观测。所以测量仪器上望远镜的放大率有一定的限度，一般为 20~45 倍。

It can be seen that the performance indicators of telescopes are mutually restricted. For example, increasing magnifying power leads to improvement in resolution and observation accuracy, but reduces the field angle and brightness, which is not conducive to observation. Therefore, the magnifying power of the telescope on measuring instruments is limited to a certain extent, generally 20-45 times.

## （2）水准器

(2) Bubble

水准器是用以置平仪器的一种设备，是测量仪器上的重要部件。水准器分为管水准器和圆水准器两种。

As an important part of measuring instruments, bubbles are a kind of device used to level the instrument. There are two types of bubbles, i.e. bubble tubes and circular bubbles.

①管水准器。

① Bubble tube.

管水准器又称水准管，是一个封闭的玻璃管，管的内壁在纵向磨成圆弧形，其半径为 0.2~100 m。管内盛酒精或乙醚或两者混合的液体，并留有一气泡（图 2-8）。管面上刻有间隔为 2 mm 的分划线，分划的中点称为水准管的零点。过零点与管内壁在纵向相切的直线称为水准管轴（*LL*）。当气泡的中心点与零点重合时，该现象称为气泡居中，气泡居中时水准管轴位于水平位置。

Bubble tubes are also called spirit levels, shaped into closed glass tubes, with the inner wall longitudinally ground into a circular arc of 0.2-100 m in radius. The tube is filled with alcohol, ether or a mixture of both, containing a bubble (Figure 2-8). The tube housing is engraved with graduations spaced 2 mm, and their midpoint is called the zero of bubble tubes. The straight line passing through the zero and longitudinally tangent to the inner wall is called the bubble tube axis (*LL*). It is called a centered bubble when its center point coincides with the zero, indicating the axis of the bubble tube is leveled.

水准管上一格（2 mm）所对应的圆心角称为水准管的分划值。根据几何关系可以看出，分划值也是气泡移动一格水准管轴所变动的角值（图 2-9）。水准仪上水准管的分划值为 10″~20″，水准管的分划值越小，视线置平的精度越高。但水准管的置平精度还与水准管的研磨质量、液体的性质和气泡的长度有关。在这些因素的综合影响下，使气泡移动 0.1 格时水准管轴所变动的角值称为水准管的灵敏度。能够被气泡的移动反映出水准管轴变动的角值越小，水准管的灵敏度就越高。

The central angle corresponding to each division of graduations (2 mm) on bubble tubes is called the scale unit. It can be seen from the geometric relationship that the scale unit is also the angle change of the bubble tube axis caused by bubble movement per division (Figure 2-9). The scale unit of the bubble tube on levels is 10″-20″, and the smaller the scale unit of the bubble tube, the higher the accuracy of line-of-sight leveling. However, the leveling accuracy of bubble tubes is also related to the grinding quality, the liquid properties, and the bubble length. Under the combined influence of these factors, the angle change of the bubble tube axis caused by bubble movement for 0.1 division is called the sensitivity of bubble tubes. The smaller the angle change caused by bubble movement, the higher the sensitivity of bubble tubes.

为了提高气泡居中的精度，在水准管的上面安装一套棱镜组（图 2-10），使两端各有半个气泡的像被反射到一起。当气泡居中时，两端气泡的像就能符合。故这种水准器又称为符合水准器。

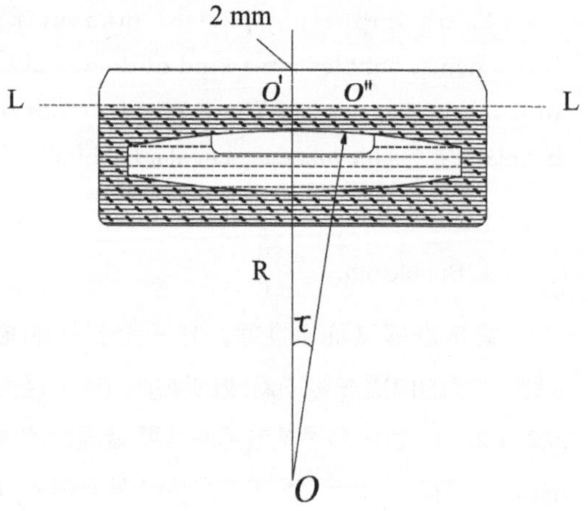


图2-8 水准管

Figure 2-8 Bubble Tube

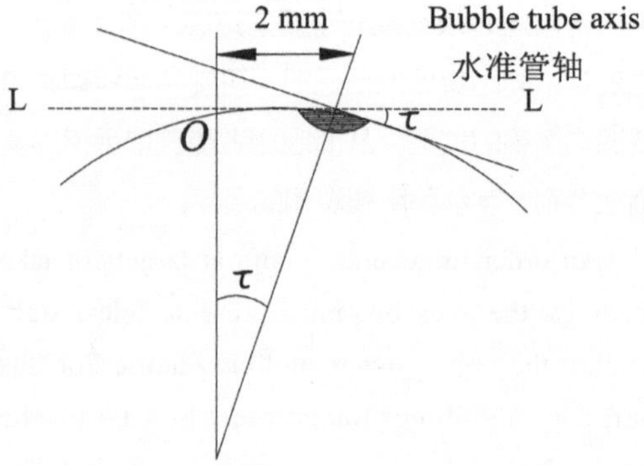


图2-9 水准管分划

Figure 2-9 Graduation of Bubble Tube

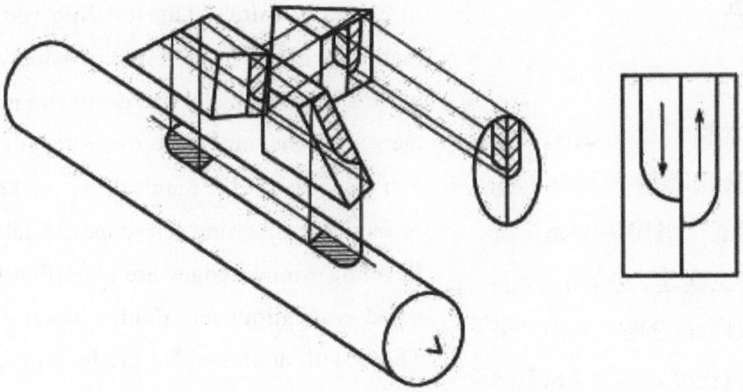

图2-10　符合水准器

Figure 2-10 Coincidence Bubble

In order to improve the accuracy of bubble centering, a set of prisms is installed on the top of bubble tubes (Figure 2-10), to reflect the images of half bubbles at both ends in coincidence. When the bubble is centered, the bubble images at both ends coincide. Therefore, this kind of bubble is also called a coincidence bubble.

②圆水准器。

② Circular bubble.

圆水准器是一个封闭的圆形玻璃容器，顶盖的内表面为一球面，半径为 0.12~0.86 m，容器内盛乙醚类液体，留有一小圆气泡（图2-11）。容器顶盖中央刻有一小圈，小圈的中心是圆水准器的零点。通过零点的球面法线是圆水准器的轴，当圆水准器的气泡居中时，圆水准器的轴位于铅垂位置。圆水准器的分划值，是顶盖球面上 2 mm 弧长所对应的圆心角值，水准仪上圆水准器的角值为 8′~15′。

Circular bubbles are shaped into a closed round glass container with a spherical inner face at the top cover and a radius of 0.12-0.86 m, and filled with ether liquid containing a small round bubble (Figure 2-11). There is a small circle engraved at the center of the top cover, and the center of the small circle is called the zero of circular bubbles. The spherical normal passing through the zero is the axis of circular bubbles, which is vertical when the bubble is centered. The central angle value ranging 8′-15′, corresponding to the arc length of 2 mm on the spherical face of the top cover, is deemed the scale unit of circular bubbles.

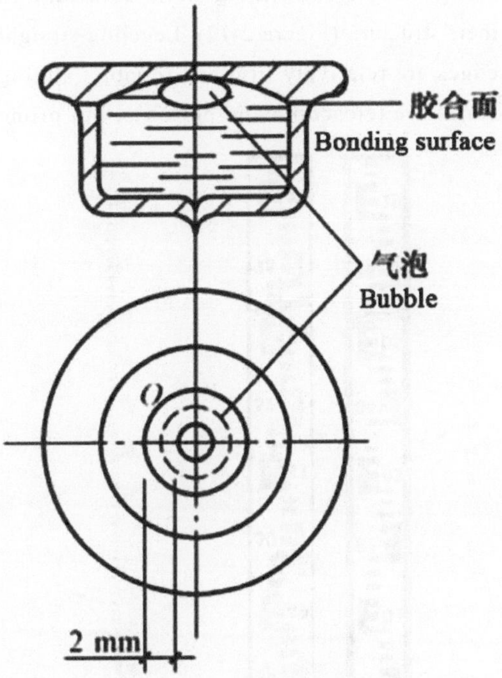


图2-11　圆水准器

Figure 2-11 Circular Bubble

## 2.2.3 水准尺和尺垫

### 2.2.3 Leveling Rod and Staff Plate

水准尺用干燥优质木材或铝合金制成，根据它们的构造可分为直尺和塔尺两种（图2-12）。直尺比较坚固可靠；塔尺能伸缩，携带方便，但接合处容易产生误差。水准尺尺面绘有 1 cm 或 5 mm 黑白相间的分格，米和分米处注有数字，尺底为零。其长度分别为 3 m 和 5 m。为了便于倒像望远镜读数，注的数字常倒写。直尺又分单面水准尺和双面水准尺两种。双面水准尺的一面为黑色的分划，另一面为红色的分划，每两根为一对。两根的黑面都以尺底为零，而红面的尺底分别为 4.687 m 和 4.787 m。在同一个测站上可同时利用黑面尺、红面尺测出两个高差进行测站检核。

Leveling rods are made of dry high-quality timber or aluminum alloy in two types of leveling straight edges and sliding staffs according to their structure (Figure 2-12). Leveling straight edges are relatively firm and reliable; sliding staffs are telescopic and portable, but prone to errors at joints. The leveling rod surface is marked with 1 cm or 5 mm divisions in alternate black and white, meter and decimetre numbers, and the zero at the bottom. Across a length of 3 m and 5 m respectively, the numbers are marked reversely to facilitate inverting telescopes to take readings. Leveling straight edges are classified into single-sided graduations and double-sided graduations. Those with double-sided graduations are marked in black on one side and in red on the other side, and are paired usually for measurement. The black side of both is marked the zero at the bottom, while the bottom of their red side is 4.687 m and 4.787 m respectively. The black and red sides are used to measure the elevation difference twice at the same observation station for checking.

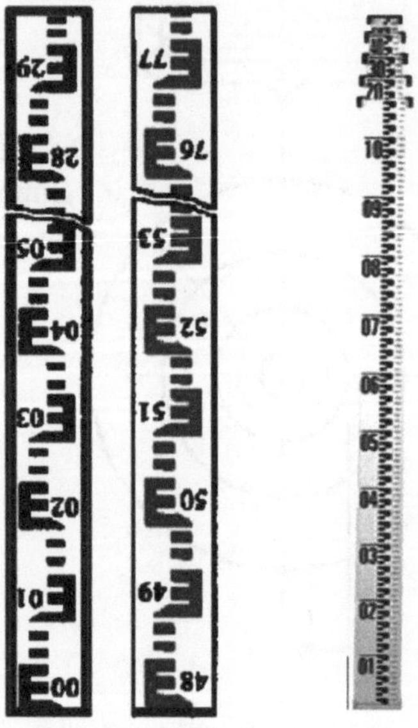

黑面
(Black side)
红面
(Red side)

图2-12　直尺、塔尺

Figure 2-12 Leveling Straight Edge, Sliding Staff

塔尺有 2 m 和 5 m 两种，由两节或三节套接而成。尺的底部为零点，尺上黑白（或红白）格相间，每格宽度为 1 cm 或 0.5 cm，每处注有数字，数字有正字和倒字两种，分米上的红色或黑色圆点表示米数。塔尺多用于地形测量、工程测量，可以缩短长度，便于携带，但接头处容易损坏，影响尺子的精度，故而常用于五等水准测量中。

Sliding staffs are in two types of 2 m and 5 m, and are composed of two or three sections socketed into a whole. They are marked with the zero at the bottom, and divisions in alternate black and white (or red and white), of 1 cm or 0.5 cm in width and accompanied with numbers each, either in normal or reversed orders. The red or black dots on decimetre numbers indicate meters. Telescopic and portable sliding staffs are mostly used for topographic surveys and engineering surveys. However, due to their joints prone to damage affecting the accuracy of observation results, they are often used in fifth-order leveling.

尺垫是用于转点上的一种工具，用钢板或铸铁制成（图 2-13）。使用时应把三个尖脚踩入土中，把水准尺立在突出的圆顶上。尺垫可使转点稳固，防止下沉。

Staff plates are a kind of tool used for erecting turning points and are made of steel plates or cast iron (Figure 2-13). Their three sharp feet are trodden into the soil before placing a leveling rod onto the protruding dome. They are used to stabilize turning points and prevent sinking.

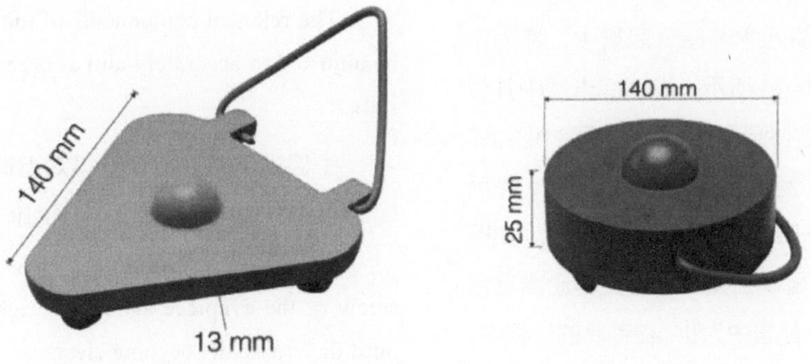


图2-13　尺垫

Figure 2-13 Staff Plate

# 2.3　水准仪的操作

## 2.3 Operation of Levels

$DS_3$ 型微倾式水准仪的操作程序分为安置仪器、粗略整平、瞄准水准尺、精确整平、读数五个步骤。

The operation procedure of $DS_3$ tilt levels is divided into five steps: placement, rough leveling, aiming at leveling rod, fine leveling and taking readings.

### 2.3.1 安置仪器

### 2.3.1 Placement

安置仪器是将仪器正确安装在可伸缩的三脚架上并置于两观测点之间。首先选好测站位置（尽量使前、后视距相等），松开三脚架伸缩螺旋，合拢三脚架并调整高度约与操作者的肩齐平，再旋紧伸缩螺旋。然后打开三脚架，将其撑稳，目测架头大致水平，踩实三脚架的三个脚尖。从仪器箱中取出仪器安置在架头上，旋紧连接螺旋。

A level is placed correctly on a telescopic tripod between two observation points as follows. First, select a position for the observation station (the foresight distance and backsight distance to be equal as far as possible), unfasten the telescopic screw of the tripod, close the tripod and adjust the height to be approximately level to the operator, and fasten the telescopic screw; then open the tripod, support it stably, visually check to ensure the tripod head is roughly leveled, and tread on the three sharp feet of the tripod fully into the soil; take out the level from the storage box and place it onto the tripod head, and fasten the connecting screw.

## 2.3.2 粗略整平

### 2.3.2 Rough Leveling

转动脚螺旋使圆水准器气泡居中，称为粗略整平。如图 2-14(a) 所示，当气泡未居中并位于 *a* 处时，可按图中所示方向用两手同时相对转动脚螺旋①和②，使气泡从 *a* 处移至 *b* 处；然后用一只手转动另一脚螺旋③，使气泡居中，如图 2-14(b) 所示。重复以上操作，直至转动望远镜在任何方向时气泡都居中为止。在整平的过程中，气泡的移动方向与左手大拇指运动的方向一致。

Turning the foot screws to center the circular bubble is called rough leveling. As shown in Figure 2-14 (a), when the bubble is not centered and located at *a* point, oppositely turn the foot screws ① and ② with both hands at the same time to move the bubble at *a* to *b*; then turn the other foot screw ③ with one hand to center the bubble, as shown in Figure 2-14 (b). The above operation is repeated until the bubble is always centered when the telescope is rotated in any direction. During this rough leveling process, the bubble moves in the same direction as the left thumb.

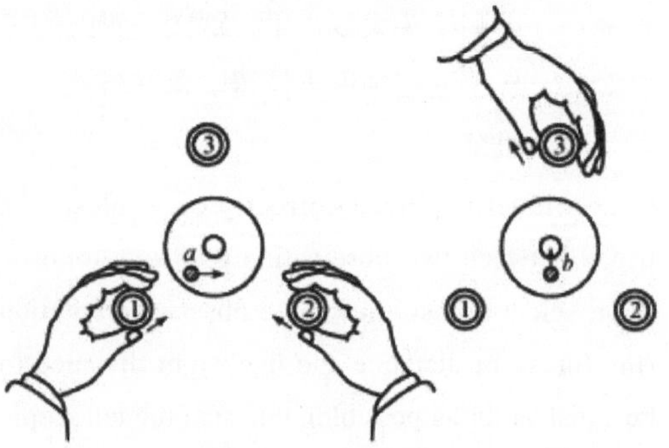


图2-14　水准仪粗略整平

Figure 2-14　Rough Leveling of Level

## 2.3.3 瞄准水准尺

### 2.3.3 Aiming at Leveling Rod

调节望远镜有关部件，达到准确瞄准目标（水准尺）的目的。

The relevant components of the telescope can be adjusted to accurately aim at the target (leveling rod).

①目镜调焦。把望远镜对着明亮的背景，转动目镜对光螺旋，使十字丝清晰。

① Eyepiece focusing. Turn the focusing screw of the eyepiece against a bright background until the crosshairs become clear.

②粗瞄目标。松开水平制动螺旋，并以制动螺旋为把手，转动望远镜，使照门和准星的连线对准水准尺，旋紧制动螺旋。

② Coarse sighting. Unfasten the horizontal clamp screw, turn the telescope with the clamp screw as a handle, to align the connecting line between the front sight and rear sight to the leveling rod, and fasten the clamp screw.

③物镜调焦。从望远镜中观察，转动物镜调焦螺旋，使水准尺成像清晰。

③ Objective lens focusing. Turn the focusing screw of the objective lens while observing through the telescope, until the leveling rod is clearly imaged.

④精确瞄准。转动水平微动螺旋，使十字丝竖丝照准水准尺中央或边缘，如图 2-15 所示。

④ Accurate sighting. Rotate the horizontal fine-motion screw to make the vertical line of crosshairs aim at the center or edge of the leveling rod, as shown in Figure 2-15.

⑤消除视差。眼睛在目镜端上、下微微移动，若发现十字丝横丝在水准尺上的读数也随之变化，则这种现象称为视差。这是因为物像与十字丝分划板平面没有重合，如图 2-16(a) 所示。视

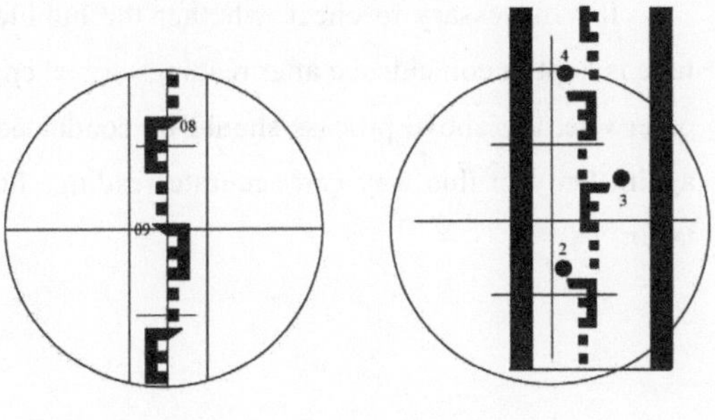

(a) (b)

图2-15　瞄准水准尺

Figure 2-15 Aiming at Leveling Rod

差会产生明显的读数误差，所以必须消除。消除视差的方法是仔细转动目镜对光螺旋与物镜调焦螺旋，直至物像与十字丝分划板平面重合，如图 2-16(b) 所示。

⑤ Eliminating parallax. Any change of the reading on the leveling rod corresponding to the horizontal line of crosshairs found while keeping an eye moving slightly up and down at the eyepiece end is called parallax. It is caused by the failure in a coincidence of the object image with the crosshair reticle, as shown in Figure 2-16 (a). Parallax directly related to obvious reading errors must be eliminated. The method to eliminate parallax is to carefully rotate the focusing screw of eyepieces and the focusing screw of objective lenses until the object image coincides with the crosshairs plane, as shown in Figure 2-16 (b).

## 2.3.4 精确整平

### 2.3.4 Fine Leveling

精确整平是调节微倾螺旋，达到水准管气泡居中、视线精确水平的目的。慢慢转动微倾螺旋，使观察窗中符合水准器气泡的影像符合。左侧影像移动的方向与右手大拇指转动方向相同。由于气泡影像移动有惯性，在转动微倾螺旋时要慢、稳、轻。

Fine leveling is to adjust the tilt screw until centering the bubble in the bubble tube and the line

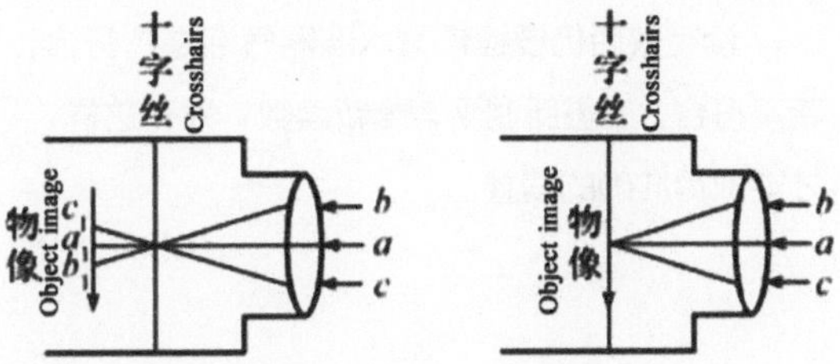


(a)存在视差　　(b)消除视差

(a) With Parallax　　(b) Without Parallax

图2-16　十字丝视差现象

Figure 2-16 Crosshair Parallax

of sight becoming accurate and horizontal. The tilt screw is slowly rotated to make the image of the coincidence bubble coincide in the observation window. The left image moves in the same direction as the right thumb. The tilt screw should be rotated slowly, steadily and lightly to overcome the inertia of bubble image movement.

## 2.3.5 读数

### 2.3.5 Taking Readings

读数时观察十字丝上的物像，达到用十字丝横丝截读水准尺上读数的目的。

The object image on the crosshairs is observed to cut out and take readings on the leveling rod with the horizontal line of crosshairs.

精确整平后，可以读取标尺读数。读数是读取十字丝中丝（横丝）截取的标尺数值。读数时，从上向下（倒像望远镜），由小到大先依次估读 mm、cm、dm、m 四位数，空位填零。为了方便，可不读小数点。

The scale readings should not be taken until accurate leveling. The scale values cut out by the middle line (horizontal line) of the crosshairs are taken as readings. Readings are taken from top to bottom (inverting telescope), estimated mm, cm, dm and m in an increasing sequence with four digits. Any null digit is filled with zero. The decimal point may be omitted for convenience.

读完数后仍要检查管水准器气泡是否符合，若不符合，应重新调平，重新读数。只有这样，才能取得准确的读数。

It is necessary to check whether the bubble tube is kept in coincidence after readings are taken; otherwise, the above process should be conducted again. Only in this way can accurate readings be taken.

# 2.4 水准测量的方法

# 2.4 Methods of Leveling

## 2.4.1 水准点

## 2.4.1 Benchmark

为统一全国的高程系统和满足各种测量的需要，测绘部门在全国各地埋设并测定了很多高程标志，这些点称为水准点 (benchmark, *BM*)。在国家高程系统中，按精度要求将测定的水准点分别称为一、二、三、四等水准点。

In order to unify the national elevation system and meet the needs of various surveys, the surveying and mapping authority has buried and measured numerous elevation marks throughout the country, which are called benchmarks (*BM*). In the national elevation system, the measured benchmarks are ranked into first, second, third and fourth-order benchmarks respectively according to the accuracy requirements.

二、三、四等水准点标志可采用磁质或金属等材料制作，其规格如图 2-17 和图 2-18 所示。三、四等水准点及四等以下高程控制点也可利用平面控制点点位标志。墙角水准点标志制作和埋设规格结构图如图 2-19 所示。

The second, third and fourth-order benchmarks are made of magnetic or metal materials, and their specifications are shown in Figure 2-17 and Figure 2-18. Horizontal control point marks may also be used as the third and fourth-order benchmarks and elevation control points below the fourth-order. The fabrication and embedding specifications and structure of benchmarks at wall corners are shown in Figure 2-19.

二、三等水准点标石规格及埋设结构图如图 2-20 所示。四等水准点标石规格及埋设结构图如图 2-21 所示。冻土地区的水准点标石规格和埋设深度，可自行设计。线路测量专用高程控制点结构可按图 2-20 所示制作，也可自行设计。

The specifications and embedding structure of the second and third-order benchmarks are shown in Figure 2-20. The specification and burial structure of the fourth-order benchmarks are shown in Figure 2-21. The specifications and burial depth of benchmarks in frozen soil areas may be customized. The special elevation control points for route surveys may be fabricated in the structure as shown in Figure 2-19 or customized.

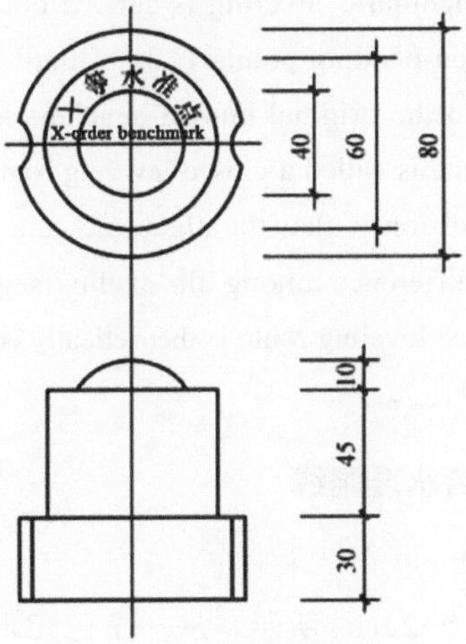


图2-17　磁质标志图

Figure 2-17  Magnetic Benchmark

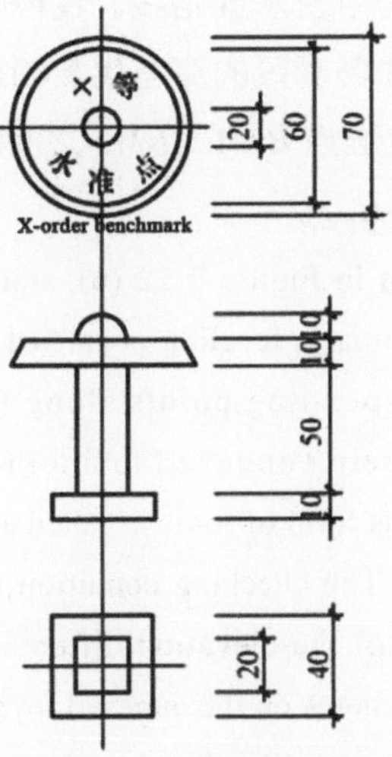


图2-18　金属标志图

Figure 2-18  Metal Benchmark

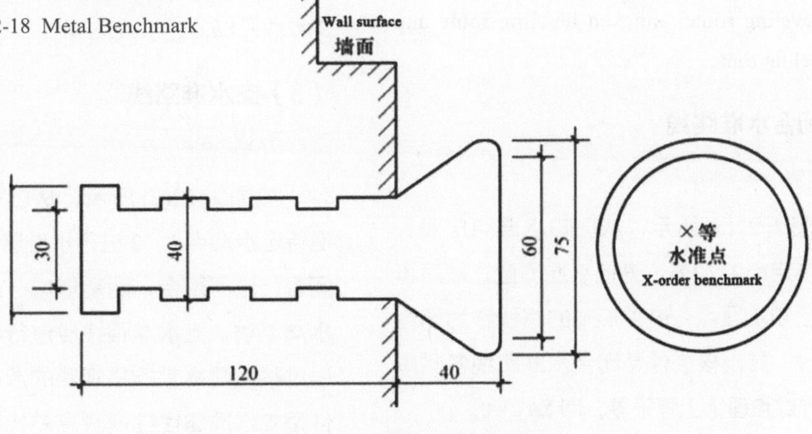


图2-19　墙角水准点标志图

Figure 2-19  Benchmark at Wall Corner

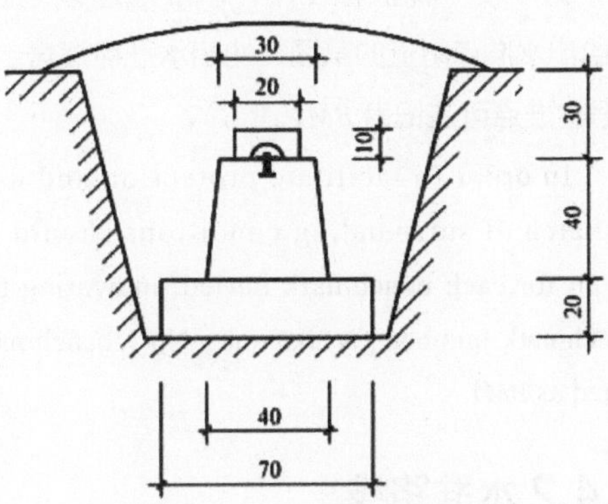


图2-20　二、三等水准点标石

Figure 2-20  Second-order and Third-order Benchmarks

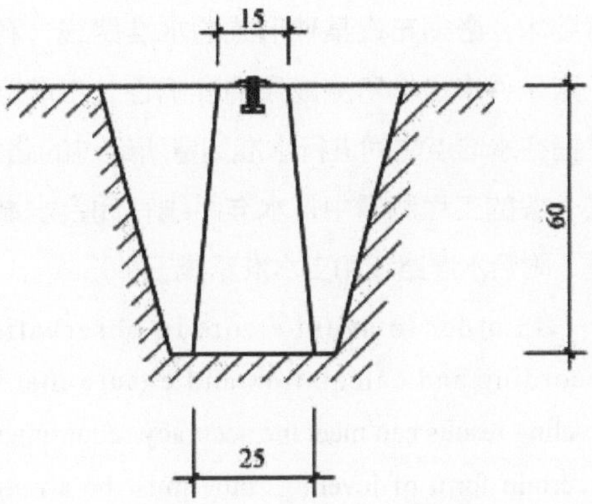


图2-21　四等水准点标石

Figure 2-21  Fourth-order Benchmark

为了便于保护和使用，水准点埋设之后，应给出水准点附近的草图，注明水准点编号，如1号水准点可注记为 $BM1$。

In order to facilitate protection and use, a sketch of surrounding conditions should be given for each benchmark buried, indicating the benchmark number, for example, No.1 benchmark noted as $BM1$.

## 2.4.2 水准路线

## 2.4.2 Leveling Route

在水准测量中，为了避免观测、记录和计算发生误差，并保证测量成果能达到一定的精度要求，必须布设某种形式的水准路线，利用一定条件来检核所测成果的准确性。水准路线是指在水准点之间进行水准测量所经过的路线。在一般的工程测量中，水准路线有闭合水准路线、附合水准路线和支水准路线三种形式。

In order to avoid errors in observation, recording and calculation and ensure that the leveling results can meet the accuracy requirements, a certain form of leveling route must be arranged to check whether observation results are correct depending on certain conditions. Leveling routes refer to the route along which leveling is carried out between benchmarks. There are three forms of leveling routes in general engineering surveys: closed leveling route, annexed leveling route and open leveling route.

### （1）闭合水准路线

### (1) Closed leveling line

如图 2-22(a) 所示，从已知水准点出发，沿高程待定点 1、2……进行水准测量，最后再回到原已知水准点，这种形式的路线称为闭合水准路线。其校核条件是闭合水准路线各测段的高差代数和理论上等于零，即 $\Sigma h_{理}=0$。

As shown in figure 2-22 (a), starting from the known benchmark, leveling is carried out along the elevation-pending points 1, 2, ..., until finally returning to the original known benchmark. This form of route is called a closed leveling route. The check condition is that the algebraic sum of the elevation difference among all leveling segments on the closed leveling route is theoretically equal to zero, i.e. $\Sigma h_{\text{Theoretical}}=0$.

### （2）附合水准路线

### (2) Annexed leveling route

如图 2-22(b) 所示，从一个已知水准点出发，沿路线上各待测高程的点进行水准测量，最后附合到另一个已知水准点上，这种水准路线称为附合水准路线。其校核条件是附合水准路线各测段的高差代数和理论上等于两个已知水准点 $BM1$、$BM2$ 之间的高差，即 $\Sigma h_{理}=H_{BM1}-H_{BM2}$。

As shown in Figure 2-22 (b), starting from a known benchmark, leveling is carried out along the elevation-pending points along the route, until finally being annexed to another known benchmark. This form of route is called an annexed leveling route. The checking condition is that the algebraic sum of the elevation difference among all leveling segments on the annexed leveling route is theoretically equal to the elevation difference between the two known benchmarks $BM1$ and $BM2$, i.e. $\Sigma h_{\text{Theoretical}}=H_{BM1}-H_{BM2}$.

### （3）支水准路线

### (3) Open leveling route

如图 2-22(c) 所示，从已知水准点出发，沿待定水准点 1、2 进行水准测量，其路线既不闭合，也不附合，而是形成一条支线，称为支水准路线。支水准路线应进行往返测量，以便通过往返测高差检核观测的准确性。其校核条件是支水准路线往返测高差代数和理论上等于零，即 $\Sigma h_{往理}+\Sigma h_{返理}=0$。

As shown in Figure 2-22 (c), starting from a known benchmark, leveling is carried out along the elevation-pending points 1, 2, ..., on a branch route, which is neither closed nor attached. This form of route is called an open leveling route. Forward and backward observations shall be carried out for an open leveling route to generate elevation differences for check of observation accuracy. The check condition is that the algebraic sum of the forward and backward observations on the closed leveling route is theoretically equal to zero, i.e.. $\Sigma h_{\text{Forward and theoretical}}+\Sigma h_{\text{Backward and theoretical}}=0$.

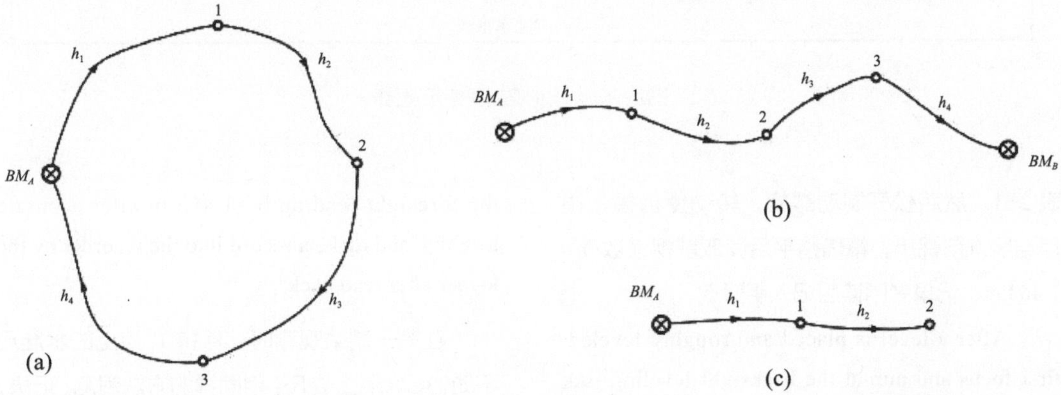


图2-22　水准路线

Figure 2-22 Leveling Route

## 2.4.3 水准测量的实施方法

### 2.4.3 Leveling Procedure

水准点埋设完毕，即可按选定的水准路线进行水准测量。如图 2-23 所示，已知水准点 $A$ 点的高程为 76. 668 m，欲测定 $B$ 点的高程。

Leveling can be carried out against the buried benchmarks and along the selected leveling route. As shown in Figure 2-21, against benchmark $A$ known with 76.668 m in elevation, the elevation of point $B$ can be measured.

如果安置一次仪器不能测出两点之间的高差，必须设置多个测站。作业时，先在适当位置选择转点 $T_1$，在水准点 $A$ 和转点 $T_1$ 上立尺，然后选择测站点安置仪器（水准仪至前、后视点的距离应尽量相等），施测第一测站。再选择转点 $T_2$ 在转点 $T_1$ 和 $T_2$ 上立尺，用同样的方法施测第二测站，依此类推，直至 $B$ 点。具体步骤如下。

If the elevation difference between the two points cannot be measured by placing an observation station, multiple stations must be set up. $A$ turning point $T_1$ is selected at an appropriate position to start the leveling operation, leveling rods are erected at the benchmark $A$ and the turning point $T_1$, and then a level is placed at the selected observation station (the distances from the level to the foresight and backsight points to be as equal as possible) to conduct the first observation. Another turning point $T_2$ is selected then, leveling rods are erected at the turning points $T_1$ and $T_2$, to conduct the second observation with the same method, and so on until point $B$. The specific steps are described as follows.

### （1）观测与记录

#### (1) Observation and recording

安置仪器与粗略整平后，首先调焦与照准后视尺，精确整平后读取后视读数 $a_1$=1.878 m，记录员回读后记入水准测量手簿相应栏内，见

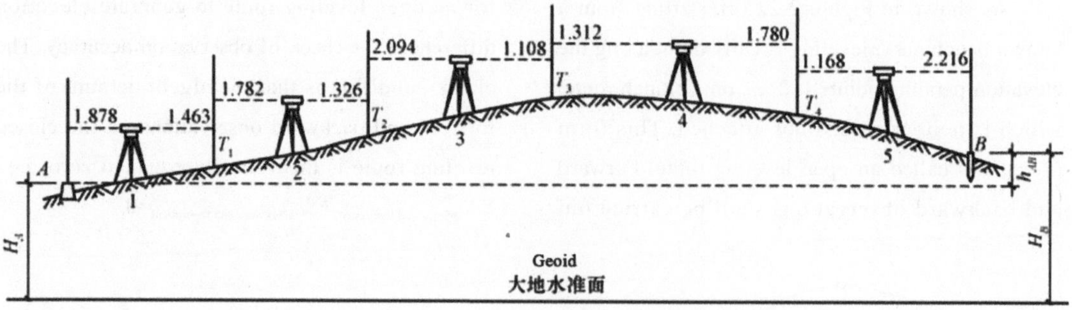


图2-23　外业观测程序示意图

Figure 2-23 Schematic Diagram of Field Observation Procedure

表 2-1。然后松开制动螺旋，转动望远镜，调焦与照准前视尺，精确整平后读取前视读数 $b_1$= 1.463 m，记录者回读后记入手簿。

After a level is placed and roughly leveled, first focus and aim at the backsight leveling rod, and then take the backsight reading $a_1$=1.878 m after accurate leveling, and make a record into the corresponding column of the leveling recorder by the logger after read back, as shown in Table 2-1; then unfasten the clamp screw, rotate the telescope, focus and aim at the foresight leveling rod, take the foresight reading $b_1$=1.463 m after accurate leveling, and make a record into the recorder by the logger after read back.

在第一测站观测时，保持 $T_1$ 点上的水准尺不动，在 $T_2$ 点上立尺，用同样的方法观测、记录，直至测到 $B$ 点。

In the first observation, it is asked to erect a leveling rod at point $T_2$ while retaining the leveling rod at point $T_1$, and make observation and recording with the same method until point $B$.

表 2-1　水准测量手簿

Table 2-1 Leveling Recorder

日期: Date:　天气: Weather:　观测者: Observer:　班组: Shift Team:　仪器: Level:　记录者: Logger:

| 测站 Observation Station | 测点 Observation Point | 后视读数 / m Backsight reading / m | 前视读数 / m Foresight reading / m | 高差 / m Elevation difference / m | 高程 / m Elevation / m |
|---|---|---|---|---|---|
| 1 | $A$ | 1.878 | | +0.415 | 76.668 |
| | $T_1$ | | 1.463 | | |
| 2 | $T_1$ | 1.782 | | +0.456 | 77.083 |
| | $T_2$ | | 1.326 | | |
| 3 | $T_2$ | 2.094 | | +0.986 | 77.539 |
| | $T_3$ | | 1.108 | | |
| 4 | $T_3$ | 1.312 | | −0.468 | 78.525 |
| | $T_4$ | | 1.780 | | |
| 5 | $T_4$ | 1.168 | | −1.048 | 78.057 |
| | $B$ | | 2.216 | | 77.009 |
| 计算校核 Calculation check | | $\sum a$=8.234 | $\sum b$=7.893 | $\sum h$= 0.341 | 77.009−76.668=0.341 |
| | | $\sum a-\sum b$=0.341 | | | |

### （2）计算和检核

### (2) Calculation and checking

①计算每一测站的高差。

① The elevation differences between observation stations are calculated as follows.

$$h_1=a_1-b_1=1.878-1.463=+0.415$$

$$h_2=a_2-b_2=1.782-1.326=+0.456$$

…

将计算出来的高差记入手簿相应栏内，见表 2-1。

The calculated elevation differences are entered into the corresponding column of the recorder, as shown in Table 2-1.

②计算 *B* 点高差。

② The elevation difference of point *B* is calculated as follows.

$$h_{AB}=\Sigma h=\Sigma a-\Sigma b=0.341\text{ m}$$

*B* 高程为

The elevation of point *B* is calculated as follows.

$$H_{AB}=H_A+h_{AB}=76.668+0.341=77.009$$

③计算校核。

③ Calculation check.

为了保证计算数据正确，需要进行计算校核。分别计算后视读数代数和减去前视读数代数和、各测站高差代数和 *A*、*B* 两点高程之差，这三个数字应相等，否则，计算有误。

In order to ensure the correctness of the calculation data, a calculation check is necessary. Calculations are carried out to generate the algebraic sum of backsight readings minus the algebraic sum of foresight readings, the algebraic sum of elevation differences among observation stations, and the elevation difference between points *A* and *B* respectively. These three results shall be equal; otherwise, the calculation is considered incorrect.

### （3）测站检核

### (3) Checking at observation stations

外业观测结束后，为了保证观测高差正确无误，必须对每测站的观测高差进行检核，这种检核称为测站检核。测站检核常采用双仪高法和双面尺法进行。

In order to ensure that the elevation differences from the field observations finished are correct, the elevation difference from each observation must be checked, which is called checking at observation stations. The observation methods with double instrument levels and with double-sided graduations are commonly used for checking at observation stations.

①双仪高法。

① Observation method with double instrument levels.

在每个测站上，利用两个不同的仪器分别观测高差，若值不超过限值 5 mm，则取平均值作为该测站的观测高差。

Elevation differences are observed with two different instrument levels respectively at each observation station. The results not exceeding the limit value of 5 mm are averaged as the observed elevation difference of the station.

②双面尺法。

② Observation method with double-sided graduations.

在每个测站上，不改变仪器高度，分别读黑面读数、红面读数，计算出黑面高差和红面高差。若两高差之差不超过限值 3 mm，则取平均值作为该测站的观测高差。

Without changing the instrument level, the readings on the black side and the red side are taken at each observation station, and the elevation difference between the black side and the red side is calculated. The two elevation differences with a difference not exceeding the tolerance of 3 mm are averaged as the observed elevation difference of the station.

# 2.5 水准测量结果的处理

## 2.5 Processing of Leveling Results

在水准测量外业工作中，无论采用哪种测量方法和测站检核，都不能保证整条水准路线的观测高差计算没有错误。因此，在计算测量结果时，要首先检查水准测量手簿中各项数据是否齐全、正确，然后计算高差闭合差，若高差闭合差符合精度要求，则调整高差闭合差，最后求出各点的高程。在计算时要注意“边计算边检核”。

No matter which methods are taken for observation and station checking in the field work of leveling, it can hardly be guaranteed that the calculation of the observed elevation differences is error-free in the whole leveling route. Therefore, it is necessary to check whether the leveling recorder is fully completed with accurate data before calculating the results. The closing error of elevation differences is first calculated. The calculated closing error of elevation differences meeting the accuracy requirements is adjusted for the final calculation of elevations at all points. The principle of “checking while calculating” shall be followed in the calculation.

水准测量结果处理的步骤应包括计算高差闭合差，计算高差闭合差的允许值，调整高差闭合差，计算待定点的高程。

The steps for processing leveling results shall include calculating the closing error of elevation differences; calculating the allowable values of the closing error of elevation differences; adjusting the closing error of elevation differences; calculating the elevation of surveyed points.

### 2.5.1 计算高差闭合差

### 2.5.1 Calculating the Closing Error of Elevation Differences

一条水准路线实际测出的高差和已知的理论高差之差称为水准路线的高差闭合差，用$f_h$表示，即

The difference between the elevation difference from actual observation and the known theoretical elevation difference on a leveling route is called the closing error of elevation differences of this leveling route, which is expressed by $f_h$.

$$f_h = \text{观测值} - \text{理论值}$$

$$f_h = \text{Observed value} - \text{Theoretical value}$$

①附合水准路线的高差闭合差为

① The closing error of elevation differences of an annexed leveling route is expressed as

$$f_h = \Sigma h - (H_{终} - H_{始}) \qquad (2.4)$$

$$f_h = \Sigma h - (H_{End} - H_{Start}) \qquad (2.4)$$

②闭合水准路线的高差闭合差为

② The closing error of elevation differences of a closed leveling route is expressed as

$$f_h=\Sigma h \qquad (2.5)$$

③支水准路线的高差闭合差为

③ The closing error of elevation differences of an open leveling route is expressed as

$$f_h=\Sigma h_{往}-\Sigma h_{返} \qquad (2.6)$$

$$f_h=\Sigma h_{Forward}-\Sigma h_{Backward} \qquad (2.6)$$

## 2.5.2 计算高差闭合差的允许值

### 2.5.2 Calculating the Allowable Values of the Closing Error of Elevation Differences

为了保证测量成果的精度，水准测量路线的高差闭合差不允许超过一定的范围，否则应重测。水准路线高差闭合差的允许范围称为高差闭合差的允许值。普通水准测量时，平地和山地的允许值按下式计算：

In order to ensure the accuracy of leveling results, the closing error of elevation differences on leveling routes is not allowed to exceed a certain range; otherwise, rework is required. The allowable range for the closing error of elevation differences on leveling routes covers the allowable values for the closing error of elevation differences. The allowable values for ordinary leveling on flat ground and mountain land are calculated according to the following formula:

平地：

Flat ground:

$$f_{允}=\pm 40\sqrt{\ } \qquad (2.7)$$

$$f_{Allowed}=\pm 40\sqrt{L} \qquad (2.7)$$

山地：

Mountain land:

$$f_{允}=\pm 12\sqrt{n} \qquad (2.8)$$

$$f_{Allowed}=\pm 12\sqrt{n} \qquad (2.8)$$

式中，$L$ 水准路线的总长度 (km)；

Where, $L$ is the total length of a leveling route (km);

$n$ 为水准路线的总测站数。

$n$ is total number of observation stations on the leveling route.

## 2.5.3 调整高差闭台差

### 2.5.3 Adjusting the Closing Error of Elevation Differences

当 $f_h$ 的绝对值小于 $f_{Allowed}$ 时，说明观测成果合格，可以进行高差闭合差的分配、高差改正。对于附合或闭合水准路线，一般按与路线长 $L$ 或测站数 $n$ 成正比的原则，将高差闭合差反号进行分配。则其高差改正数 $V_i$ 的计算公式为

An absolute value of $f_h$ less than $f_{Allowed}$ indicates that the observation results are conforming, enabling the assignment of the closing error of elevation differences and correction of elevation differences. For an annexed or closed leveling route, the closing error of elevation differences is generally assigned with reverse signs and according to the principle of being proportional to the route length L or the number of observation stations n. Therefore, the calculation formula for the elevation difference correction $V_i$ as follows

$$V_i=-\frac{f_h}{L}L_i \text{ 或 } V_i=-\frac{f_h}{n}n_i \qquad (2.9)$$

$$V_i=-\frac{f_h}{L}L_i \text{ or } V_i=-\frac{f_h}{n}n_i \qquad (2.9)$$

改正后的高差为

The corrected elevation difference is

$$f_{i改}=h_{i测}+V_i \quad (2.10)$$

$$f_{iCorrected}=h_{iObserved}+V_i \quad (2.10)$$

对于支水准路线，采用往测高差减去返测高差后取平均值，作为改正后往测方向的高差，即

For an open leveling route, the elevation differences in the forward observation minus the elevation differences in the backward observation are averaged as the corrected elevation difference for the forward observation.

$$h_i=(h_{往}+h_{返})/2 \quad (2.11)$$

$$h_i=(h_{Forward}+h_{Backward})/2 \quad (2.11)$$

### 2.5.4 计算待定点的高程

### 2.5.4 Calculating the Elevation of Surveyed Points

对于附合水准路线或闭合水准路线，须根据起点的已知高程加上各段调整后的高差依次推算各所求点的高程，即

For an annexed leveling route or closed leveling route, the elevation of each surveyed point shall be calculated according to the known elevation of the starting point plus the adjusted elevation difference of each section in turn, i.e.

$$H_i=H_{i-1}+h_i \quad (i=1, 2, 3, \ldots) \quad (2.12)$$

推算到终点已知的高程点上时，应与该点的已知高程相等。否则，说明计算有误，应找出原因，并重新计算。

The calculated elevation of the ending point shall be equal to the known elevation of the point. Otherwise, the calculation is deemed incorrect, and recalculation shall be conducted after the cause is identified.

对于支水准路线来说，因无法检核，故在计算中要仔细认真，确认计算无误后方能使用计算成果。

Since the calculation results of an open leveling route cannot be checked, the calculation results shall not be used until it is confirmed that the calculation is correct.

# 本章小结

# Chapter Summary

本章主要讲述了水准测量的原理、水准测量仪器及其使用、水准测量的施测方法、水准测量结果计算等内容。

This chapter mainly introduces the principle of leveling, leveling instruments and their operation, leveling procedures, leveling results calculation, etc.

1. 水准测量是利用水准仪提供的水平视线，根据水准仪在两点竖立的水准尺上的读数，先求得两点之间的高差，然后根据已知点的高程推算出未知点的高程。

1.Taking advantage of the horizontal sight provided by a level, leveling is to calculate the elevation difference between two points according to the readings on the leveling rods erected at the two points, and then calculate the elevation of the surveyed point according to the point with a known elevation.

2. 水准测量所使用的仪器为水准仪，工具为水准尺和尺垫。要在认识水准仪基本构造的基础上，重点掌握使用 $DS_3$ 型水准仪进行粗略整平的方法以及瞄准水准尺、精确整平和读数的方法。在了解水准仪应满足的几何条件的基础上，掌握圆水准器、十字丝分划板、水准管轴的检验与校正方法。

2.Leveling is carried out with levels, with the help of the tools of leveling rods and staff plates. This chapter focuses on the rough leveling, sighting, accurate leveling and reading methods of the $DS_3$ level on the basis of the basic structure of levels; the inspection and calibration methods of circular bubbles, crosshair reticles and bubble tube axis on the basis of the necessary geometric conditions of levels.

3. 水准测量的实测应重点学习观测的基本步骤、数据记录计算和测量检核这三个环节，要求掌握水准仪的高差闭合差的计算与调整。

3.For the actual measurement of leveling, emphasis is given to the basic steps of observation, data recording and calculation, result check, as well as the calculation and adjustment of the closing error elevation differences of levels.

4. 在了解水准测量误差的主要来源的基础上，掌握消除或减小误差的基本措施，提高测量精度。

4.The basic measures are given to eliminate or reduce errors and improve observation accuracy on the basis of understanding the main sources of leveling errors.

# 复习思考题

## Review Questions

## 一、填空题

### I. Fill in the Blanks

1. 水准仪按其精度，分为__________、__________、__________、__________四个等级。

1. By accuracy, levels are classified into four levels of __________, __________, __________, and__________.

2. 水准仪主要由__________、__________及__________三部分组成。

2. A level is mainly composed of three parts of__________,__________, and__________.

3. 水准仪的使用主要包括__________、__________、__________、__________、__________等基本操作步骤。

3. The operation of a level mainly includes the basic steps of __________, __________, __________, __________, and__________.

4. 根据水准测量原理，微倾式水准仪有__________、__________、__________和__________四条轴线。

4. According to the leveling principle, a tilt level comprises and four axes of __________, __________, __________, and __________.

5. 水准仪的圆水准器轴应__________于竖轴。

5. The circular bubble axis of a level shall__________the vertical axis.

## 二、选择题

## II. Choice Questions

1. 往返水准路线高差平均值的正负号是以（ ）的符号为准。

1.The sign of the average elevation difference on the forward and backward leveling routes is taken as plus or minus according to ( ).

A. 往测高差

A. elevation differences in forward leveling

B. 返测高差

B. elevation differences in backward leveling

C. 往返测高差的代数和

C. algebraic sum of elevation differences in forward and backward routes

D. 以上三者都不正确

D. none of the above

2. 已知 $A$ 点高程 $H_A$ = 62.118 m，水准仪观测到的 $A$ 点标尺的读数 $a$ = 1.345 m，则仪器视线高程为（ ）m。

2.Given that the elevation of point $A$ is $H_A$ = 62.118 m and the reading on the staff of point $A$ observed by the level is $a$ = 1.345 m, the elevation of sight of the level is ( ) m.

A. 60.773 B. 63.463 C. 62.118 D. 64.213

3. 转动目镜对光螺旋的目的是（ ）。

3.The focusing screw of the eyepiece is turned to ( ).

A. 看清远处目标

A. see a target at a long distance

B. 看清近处目标

B. see a target at a short distance clearly

C. 看清十字丝

C. see the crosshairs clearly

D. 消除视差

D. eliminate parallax

4.$DS_1$ 型水准仪的观测精度要（ ）$DS_3$ 型水准仪。

4.The observation accuracy of $DS_1$ level is (  ) $DS_3$ level.

A. 高于 B. 接近于 C. 低于 D. 等于

A. higher than B. close to C. lower than D. equal to

5. 当圆水准器气泡居中时，圆水准器轴处于（  ）位置。

5.When the bubble of the circular bubble is centered, the circular bubble axis is（  ）.

A. 竖直 B. 水平 C. 倾斜 D. 任意

A. vertical B. horizontal C. tilt D. at any position

6. 下列不属于观测误差的是（  ）。

6.Which of the following is not an observation error (  ).

A. 水准管气泡居中误差 B. 水准尺的误差

A. bubble centering error of bubble tube B. error of leveling rod

C. 读数误差 D. 水准尺倾斜误差

C. reading error D. tilt error of leveling rod

## 三、简答题

## III. Short Answer Questions

1. 简述水准仪中望远镜各组成部分的作用。

1.Briefly describe the role of all components of the telescope in levels.

2. 进行水准测量时，为什么要求前、后视距大致相等？

2.Why are the foresight and backsight distances required to be approximately equal in leveling?

3. 水准测量中的检核内容有哪些？各项检核的方法是什么？

3.Which items need to be checked in leveling? How to check for each item?

4. 什么是水准路线？什么是高差闭合差？如何计算容许的高差闭合差？

4.What is a leveling route? What is a closing error of elevation differences? How to calculate the allowable closing error of elevation differences?

[illegible] The orientation accuracy of [illegible] level.

A. [illegible] B. [illegible] C. [illegible] D. [illegible]

A. [illegible] B. [illegible] C. [illegible] D. [illegible]

[illegible]

[illegible] axis is ( )

A. [illegible] B. [illegible] C. [illegible] D. [illegible]

A. vertical B. horizontal C. [illegible] D. in any position

[illegible]

Which of the following is not [illegible] ( )

A. [illegible] B. [illegible]

A. adjusting the focusing screw of bubble tube B. [illegible] of leveling rod

C. [illegible] D. [illegible]

C. [illegible] D. [illegible] of leveling rod

三、简答题

[illegible]

[illegible] the components in the [illegible] levels.

[illegible]

2. Why are the foresight and backsight distances required to be approximately equal in leveling?

[illegible]

3. Which [illegible] in leveling? How to check [illegible]?

[illegible]

4. What is [illegible] error? What [illegible] difference? How to calculate [illegible] elevation [illegible]

建筑工程测量

ARCHITECTURAL ENGINEERING SURVEYING

3

# 角度测量

# Angle Measurement

**学习目标：**

**Learning Objectives:**

通过本章的学习，了解经纬仪的构造及其各部分的作用；熟悉水平角、竖直角测量的原理；掌握光学经纬仪的使用方法、光学经纬仪的检验步骤与校正方法、水平角和竖直角的观测方法。

Through the study of this chapiter, to learn about the structure of the theodolite and the functions of each part; to get familiar with the principles of horizontal and vertical angle measurement; to master the use methods, inspection steps and correction methods of the optical theodolite, and the observation methods of horizontal and vertical angles.

**能力目标：**

**Ability Objectives：**

能够熟练操作光学经纬仪，具有对经纬仪进行检验及简单校正的能力；能够运用测回法进行水平角和竖直角观测；能进行角度数据的计算与处理。

To be able to skillfully operate the optical theodolite, and perform simple inspection and correction of the theodolite; to be able to use the observation set method to observe horizontal and vertical angles; to be able to calculate and process angle data.

# 3.1 角度测量的原理

## 3.1 Principles of Angle Measurement

角度测量是测量的基本工作之一。角度测量是测量地面点连线的水平夹角及视线方向与水平面的竖直角。水平角测量用于计算点的平面位置；竖直角测量用于测定高差或将倾斜距离改成水平距离。角度测量常用的仪器是经纬仪。

Angle measurement is one of the basic tasks of measurement. Angle measurement is to measure the horizontal angle between the connecting lines of points on the ground and the vertical angle between the line of sight and the horizontal plane. Horizontal angle measurement is used to calculate the plane position of a point; vertical angle measurement is used to determine the height difference or convert the slant distance to horizontal distance. The commonly used instrument for angle measurement is theodolite.

### 3.1.1 水平角的测量原理

### 3.1.1 Principle of Horizontal Angle Measurement

水平角是地面上某一点到两个目标点的方向线垂直投影在水平面上的夹角，即通过这两条方向线所作两竖直面间的二面角，用$\beta$来表示，其角值范围为 0°~360°。

The horizontal angle is the included angle between the two direction lines from a certain point on the ground to two target points vertically projected on the horizontal plane, that is, the dihedral angle between the two vertical planes made through the two direction lines, expressed by $\beta$, ranging from 0° to 360°.

如图 3-1 所示，$A$ 点、$B$ 点、$C$ 点是地面上任意三个点，$AB$ 和 $AC$ 两条方向线所夹的水平角，就是通过 $AB$、$AC$ 沿两个竖直面投影在水平面 $P$ 上的两条水平线 $ab$ 和 $ac$ 的夹角 $\beta$（$\angle bac$）。

As shown in Figure 3-1, points $A$, $B$ and $C$ are three arbitrary points on the ground, and the horizontal angle between the two direction lines $AB$ and $AC$ is the angle $\beta$（$\angle bac$）between the two horizontal lines ab and ac projected on the horizontal plane P through $AB$ and $AC$.

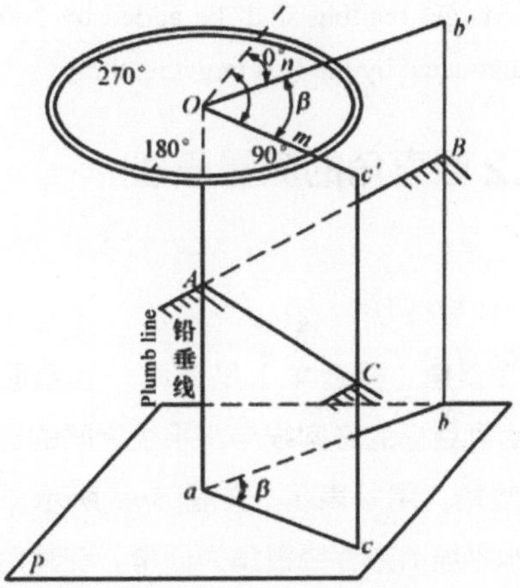


图3-1　水平角的测量原理

Figure 3-1 Principle of Horizontal Angle Measurement

为了测定水平角的大小，设想在 $A$ 点的铅垂线上任一处，水平安置一个带有顺时针均匀刻划的水平度盘，通过左方向 $AB$ 和右方向 $AC$ 的竖直面与水平度盘相交，在度盘上截取相应的读数 $a_1$ 和 $b_1$，则水平角 $\beta$ 为右方向读数 $b_1$ 减去左方向读数 $a_1$，即

To determine the size of the horizontal angle, assume that a horizontal circle with even clockwise graduation is set horizontally at an arbitrary point on the plumb line of point $A$, allow the vertical plane passing through the left direction *AB* and the right direction *AC* to intersect with the horizontal circle, and take the corresponding readings $a_1$ and $b_1$ on the dial; so, the horizontal angle $\beta$ is the reading $b_1$ in the right direction minus the reading $a_1$ in the left direction, i.e.

$$\beta = b_1 - a_1 \tag{3.1}$$

需要注意的是，水平角的角值范围为0°～360°。当第二目标读数小于第一目标读数时，应在第二目标读数值上加上360°后再减第一目标读数。

It should be noted that the horizontal angle ranges from 0° to 360°. When the second target reading is less than the first target reading, the second target reading shall be added by 360° and then subtracted by the first target reading.

### 3.1.2 竖直角的测量原理

### 3.1.2 Principle of Vertical Angle Measurement

竖直角（垂直角）是在同一竖直面内，某一点到目标的方向线与水平线之间的夹角，又称倾角，用 $\alpha$ 表示。如图3-2所示，方向线在水平线上方，竖直角为仰角，在其角值前加"+"号；方向线在水平线下方，竖直角为俯角，在其角值前加"-"号。竖直角的角值

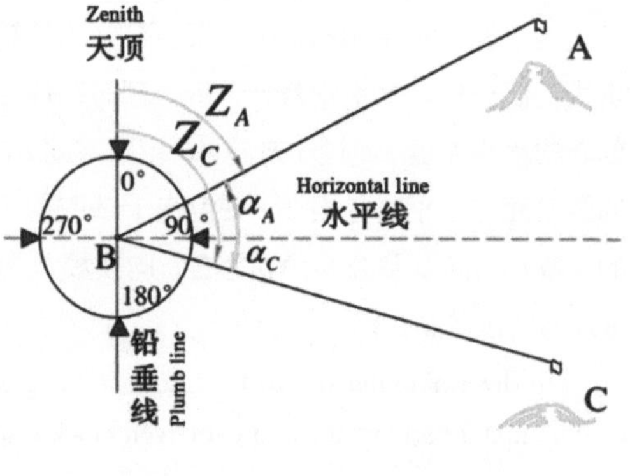


图3-2 竖直角的测量原理

Figure 3-2 Principle of Vertical Angle Measurement

范围为 -90°～+90°。

The vertical angle (perpendicular angle) is the angle between the direction line from a certain point to a target point and the horizontal line in the same vertical plane, also known as the inclination angle, expressed by $\alpha$. As shown in Figure 3-2, when the direction line is above the horizontal line, the vertical angle is the elevation angle with "+" added before its angle value; when the direction line is below the horizontal line, the vertical angle is the depression angle with "–" added before its angle value. The vertical angle ranges from –90° to +90°.

同水平角一样，竖直角的角值也是度盘上两个方向的读数之差，所不同的是，该度盘是竖直放置的，因此称为竖直度盘。望远镜瞄准目标时，倾斜视线在竖直度盘上的对应读数与水平视线的应有度数（盘左90°角，盘右270°）之差即为竖直角的角值。由于竖直角的两个方向中的一个方向是水平方向，无论对哪一种经纬仪来说，视线水平时的竖盘读数都应是90°的倍数。所以，当测量竖直角时，只要瞄准目标读出竖盘读数，即可计算出竖直角的度数。

Same as the horizontal angle, the vertical angle is also the difference between the readings in two directions on the circle. The difference is that the circle is placed vertically, so it is called a vertical circle. When the telescope is aiming at the target, the difference between the corresponding reading of the oblique line of sight on the vertical circle and the due degrees of the horizontal line of sight (90° on the left and 270° on the right of the circle) is the angle value of the vertical angle. Since one of the two directions of the vertical angle is the horizontal direction, the vertical circle reading shall be a multiple of 90° regardless of the theodolite type. Therefore, the vertical angle can be calculated by just aiming at the target and reading the vertical circle during the measurement.

# 3.2 光学经纬仪的构造及使用

# 3.2 Structure and Use of Optical Theodolite

## 3.2.1 光学经纬仪的构造

## 3.2.1 Structure of Optical Theodolite

经纬仪是测量角度的仪器，有光学经纬仪和电子经纬仪两大类。我国将经纬仪分为 $DJ_{07}$、$DJ_1$、$DJ_2$、$DJ_6$ 等不同级别。其中，D 和 J 分别是“大地测量”“经纬仪”两个汉语拼音第一个字母，数字“07”“1”“2”“6”表示该级别仪器所能达到的测量精度指标（数字表示此精度级别的经纬仪一测回方向观测中误差的秒值），数字越大，级别越低。目前，建筑工程中使用较多的光学经纬仪是 $DJ_2$ 型经纬仪和 $DJ_6$ 型经纬仪。

Theodolite is an instrument for measuring angles, including optical theodolite and electronic theodolite. Theodolites are graded as $DJ_{07}$, $DJ_1$, $DL_2$, $DJ_6$, etc. in China. Where, “D” stands for “Dadiceliang (Geodetic Survey)” and “J” for “Jingweiyi (Theodolite)”, and the numbers “07”, “1”, “2” and “6” represent the measurement precision index that can be achieved by the instrument at the corresponding grade. The numbers represent the value (in seconds) of the error in the observation direction of the theodolite at this precision grade, and the higher the number, the lower the grade. At present, the optical theodolites commonly used in construction projects are $DJ_2$ theodolites and $DJ_6$ theodolites.

### （1）$DJ_6$ 型光学经纬仪的构造

### (1) Structure of $DJ_6$ optical theodolite

$DJ_6$ 型光学经纬仪主要由照准部、水平度盘和基座三大部分组成。其基本构造如图 3-3 所示。

$DJ_6$ optical theodolite is mainly composed of three parts: alidade, horizontal circle and base. The basic structure is shown in Figure 3-3.

①照准部。

① Alidade.

照准部是指位于水平度盘之上，能绕其旋转轴旋转的部分的总称。照准部由望远镜、竖盘装置、读数显微镜、水准管、光学对中器、照准部制动螺旋和微动螺旋、望远镜制动螺旋和微动螺旋等部分组成。照准部旋转所绕的几何中心线称为经纬仪的竖轴，通常也将其旋转轴称为竖轴。照准部制动螺旋和微动螺旋用于控制照准部的转动。

Alidade is the general term of the part located on the horizontal circle that can rotate around its rotation axis. It consists of telescope, vertical circle, reading microscope, level tube, optical centering device, alidade clamp and fine-motion screws, telescope clamp and fine-motion screws, etc. The geometric centerline around which the alidade rotates is called the vertical axis of the theodolite, and its own rotation axis is usually called the vertical axis as well. The alidade clamp and fine-motion screws are used to control the rotation of the alidade.

经纬仪的望远镜构造与水准仪望远镜相同，它与横轴连在一起，当望远镜绕横轴旋转时，视线可扫出一个竖直面。望远镜制动螺旋用来控制望远镜在竖直方向上的转动，望远镜微动螺旋是当望远镜制动螺旋拧紧后，用此螺旋使望远镜在竖直方向上做微小转动，以便精确对准目标。照准部制动螺旋控制照准部在水平方向的转动。当照准部制动螺旋拧紧后，可利用

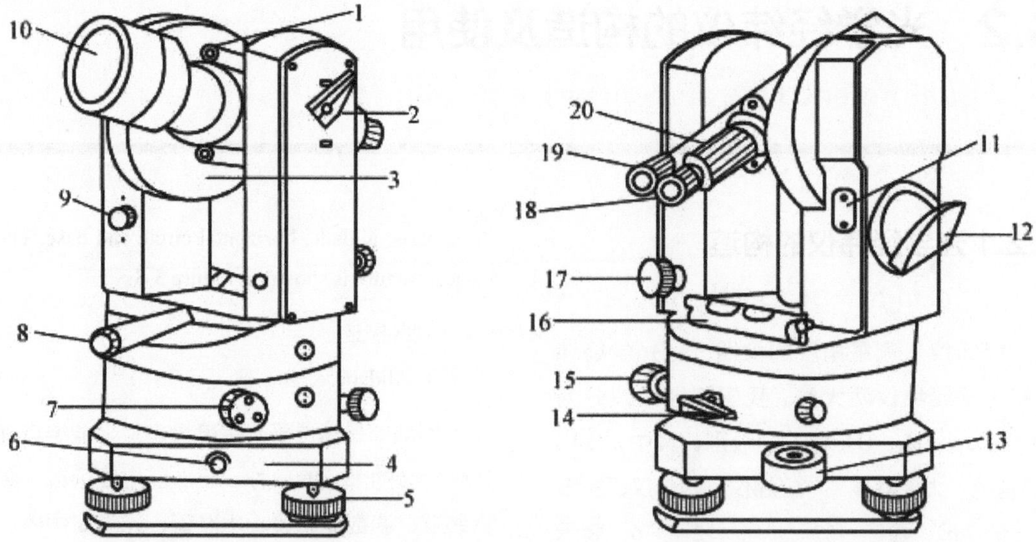


1—粗瞄器；2—望远镜制动螺旋；3—竖盘；4—基座；5—脚螺旋；6—固定螺旋；7—度盘变换手轮；8—光学对中器；9—自动归零旋钮；10—望远镜物镜；11—指标差调位盖板；12—反光镜；13—圆水准器；14—水平制动螺旋；15—水平微动螺旋；16—照准部水准管；17—望远镜微动螺旋；18—望远镜目镜；19—读数显微镜；20—对光螺旋。

1-coarse sighting device; 2-telescope clamp screw; 3-vertical circle; 4-base; 5-foot screw; 6-fixed screw; 7-circle shift handwheel; 8-optical centering device; 9-automatic zeroing knob; 10-telescope objective lens; 11-index difference adjustment cover plate; 12-reflector; 13-circular level; 14-horizontal clamp screw; 15-horizontal fine-motion screw; 16-alidade level tube; 17-telescope fine-motion screw; 18-telescope eyepiece; 19-reading microscope; 20-focusing screw.

图3-3 $DJ_6$型光学经纬仪

Figure 3-3 $DJ_6$ Optical Theodolite

照准部微动螺旋使照准部在水平方向上做微小转动，以便精确对准目标。

The structure of the theodolite telescope is the same as that of the level telescope, which is connected with the horizontal axis. When the telescope rotates around the horizontal axis, the line of sight can sweep out a vertical plane. The telescope clamp screw is used to control the rotation of the telescope in the vertical direction, and the telescope fine-motion screw is used to rotate the telescope slightly in the vertical direction when the telescope clamp screw is tightened, so as to accurately aim at the target. The alidade clamp screw is used to control the rotation of the alidade in the horizontal direction, and the alidade fine-motion screw is used to rotate the alidade slightly in the horizontal direction when the alidade clamp screw is tightened, so as to accurately aim at the target.

照准部上安装有水准管，其作用是精确整平仪器，使仪器的竖轴处于铅垂位置，并根据仪器内部应具备的几何关系使水平度盘和横轴处于水平位置。照准部上还设有光学对中器，用于光学对中。

A level tube is installed on the alidade, and it is used to accurately level the instrument, making the vertical axis of the instrument in a plumb position, and the horizontal circle and horizontal axis in the horizontal position according to the required geometric relationship inside the instrument. An optical centering device is also provided on the alidade for optical centering.

照准部上反光镜的作用是将外部光线反射进入仪器，通过一系列透镜和棱镜，将度盘和分微尺的影像反映到读数显微镜内，以便读出水平度盘或竖直度盘的读数。

The function of the reflector on the alidade is to reflect the external light into the instrument, and reflect the image of the circle and micrometer into the reading microscope through a series of lenses and prisms, so as to take the reading of the horizontal circle or vertical circle.

②水平度盘。

② Horizontal circle.

水平度盘是由光学玻璃制成的带有刻划和注记的圆盘，安装在仪器竖轴上，在度盘的边缘按顺时针方向均匀刻划成 360 份，每一份就是 1°，并注记度数。在测角过程中，水平度盘和照准部分离，不随照准部一起转动。当望远镜照准不同方向的目标时，移动的读数指标线便可在固定不动的度盘上读出不同的度盘读数，即方向值。如需要变换度盘位置，可利用仪器上的度盘变换手轮，把度盘变换到需要的读数上。

The horizontal circle is a disc made of optical glass with graduations and annotations, which is installed on the vertical axis of the instrument. The edge of the circle is evenly graduated into 360 parts clockwise, each of which is 1°, with corresponding degrees annotated. During angle measurement, the horizontal circle is separated from the alidade and does not rotate together with the alidade. When the telescope is aiming at targets in different directions, the moving reading index line can read different readings on a fixed circle, that is, the direction values. To change the position of the circle, you can use the circle shift handwheel on the instrument to shift the circle to the required reading.

③基座。

③ Base.

基座上有三个脚螺旋、一个圆水准器气泡，用来粗略整平仪器。水平度盘旋转轴套在竖轴套外围，拧紧轴套固定螺钉，可将经纬仪固定在基座上；旋松该螺钉，可将经纬仪水平度盘连同照准部从基座中拔出，但平时应将该螺钉拧紧。

There are three foot screws and one circular level bubble on the base for rough leveling of the instrument. The axis sleeve of the rotation axis of the horizontal circle is on the periphery of the vertical axis. The theodolite can be fixed on the base by tightening the sleeve fixing screw. The theodolite can be removed with alidade from the base by loosening the same screw. However, the screw should be tightened normally.

## 3.2.2 光学经纬仪的使用

## 3.2.2 Use of Optical Theodolite

### （1）$DJ_6$ 型光学经纬仪的使用

### (1) Use of $DJ_6$ optical theodolite

经纬仪的使用包括安置仪器、瞄准目标和读数三项基本操作。

The use of the theodolite includes three basic operations: placing the instrument, aiming at the target and reading.

①安置仪器。安置仪器是将经纬仪安置在测站点上，包括对中经纬仪安置和整平两项内容。

① Placing the instrument. Placing instruments is to install the theodolite on the observation station, including the placing and leveling of the theodolite.

a. 对中。对中的目的是使仪器中心与测站点标志中心位于同一铅垂线上。其具体做法如下。

a.Centering. The purpose of centering is to make the center of the instrument and the center of the observation station mark on the same plumb line. Specific steps are as follows.

第一，先松开三脚架架脚三个固定螺旋，按观测者身高调整好脚架的高度，然后将三个螺旋拧紧。

Firstly, loosen the three fixing screws of the tripod feet, adjust the length of the tripod according to the height of the observer, and then tighten the three screws.

第二，张开三脚架，将其安置在测站上，使架头大致水平，且架头中心与测站点位于同一铅垂线上。

Secondly, unfold the tripod and place it on the observation station in such a manner that the head is approximately horizontal and the center of the head is on the same plumb line as the observation station.

第三，从仪器箱中取出经纬仪放置在三脚架架头上，并使仪器基座中心基本对齐三脚架架头的中心，旋紧连接螺旋后，即可进行对中整平操作。

Thirdly, take out the theodolite from the instrument box, place it on the tripod head, and align the center of the instrument base with the center of the tripod head. After tightening the connecting screw, carry out the centering and leveling operations.

可以使用垂球对中或光学对中器对中进行经纬仪安置操作。

Use a plumb bob or optical centering device to place the theodolite.

第一，使用垂球对中法安置经纬仪。将垂球挂在连接螺旋中心的挂钩上，调整垂球线长度，使垂球尖略高于测站点。

Firstly, place the theodolite with plumb bob centering method. Hang the plumb bob on the hook at the center of the connecting screw, and adjust the length of the plumb-bob line to make the plumb bob tip slightly higher than the observation station.

初步对中。如果相差太大，可前后左右摆动三脚架架脚，或整体移动三脚架，使垂球尖大致对准测站点标志，并注意架头基本保持水平，然后将三脚架的脚尖踩入土中。

Preliminary centering. If the difference is too large, swing the legs of the tripod back and forth, or move the tripod as a whole to make the plumb bob tip roughly aligned with the observation station mark, pay attention to keeping the tripod head basically horizontal, and then step the foot tip of the tripod into the soil.

精对中。稍微旋松连接螺旋，双手扶住仪器基座，在架头上平移仪器，使垂球尖精确对中测站点标志中心，再旋紧连接螺旋。垂球对中的误差应小于 3 mm。

Precise centering. Slightly loosen the connecting screw, hold the instrument base with both hands, translate the instrument on the tripod head to make the plumb ball tip accurately aligned with the center of the observation station mark, and then tighten the connecting screw. The error of plumb bob centering shall be less than 3 mm.

精确整平。旋转脚螺旋，使圆水准器气泡居中；转动照准部，旋转脚螺旋，使管水准器气泡在相互垂直的两个方向上居中。旋转脚螺旋精确整平仪器时，注意不要破坏已完成的垂球对中关系。

Accurate leveling. Rotate the foot screw to center the circular level bubble; rotate the alidade, and turn the foot screw to center the level bubble in two directions perpendicular to each other. Be careful not to damage the centered position relation when rotating the foot screw for exact leveling of the instrument.

第二，使用光学对中器对中法安置经纬仪。光学对中器对中是利用几何光学原理，移动三脚架中的任意两脚或整个仪器在架头上平移，使光学对中器小圆圈中心对准测站点标志中心，达到对中的目的。用光学对中器对中的误差一般可控制在 3 mm 以内。

Secondly, place the theodolite with optical centering method. Optical centering device uses the principles of geometric optics for centering. Move any two feet of the tripod or the whole instrument on the tripod head to align the center of the small circle of the optical centering device with the center of the observation station mark, so as to achieve the purpose of centering. The centering error with an optical centering device can be generally controlled within 3 mm.

由于光学对中器的视线与仪器竖轴重合，只有在仪器整平后，视线才处于铅垂位置。对中时，最好也先用垂球尖大致对中，再调节对中器的目镜和物镜，使分划板小圆圈和测站点标志同时清晰，然后固定一个架脚，移动其余两个架脚，使照准圈大致对准测站点标志，并踩踏三脚架，使其稳固地插入地面。若对中偏离较小，也可稍旋松连接螺旋，两手扶住仪器基座，在架头上平移仪器，使目镜分划板小圆圈中心精确对中测站点标志中心，最后旋紧连接螺旋。

Since the line of sight of the optical centering device coincides with the vertical axis of the instrument, the line of sight is only in the vertical position after the instrument is leveled. During centering, it is better to first roughly center the instrument with the plumb bob tip, and then adjust the eyepiece and objective lens of the centering device to make the small circle of the reticle and the mark of the observation station clear at the same time; next, fix one leg and move the other two legs to make the sighting circle roughly aligned with the mark of the observation station, and step on the tripod to firmly insert it into the ground. If the centering deviation is small, slightly loosen the connecting screw, hold the instrument base with both hands, translate the instrument on the tripod head, accurately align the center of the small circle of the eyepiece reticle with the center of the observation station, and finally tighten the connecting screw.

b. 整平。整平的目的是使仪器竖轴处于铅垂位置，水平度盘处于水平位置。其具体做法如下。

b. Leveling. The purpose of leveling is to make the vertical axis of the instrument in the vertical position and the horizontal circle in the horizontal position. Specific steps are as follows.

第一，粗略整平。根据圆水准器气泡偏离情况，分别伸长或缩短三脚架架脚，使圆水准器气泡居中。

Firstly, rough leveling. According to the bubble deviation of the circular level, extend or shorten the legs of the tripod to center the bubble of the circular level.

第二，精确整平。如图 3-4（a）所示，精确整平时，先转动仪器的照准部，使照准部水准管与任一对脚螺旋的连线平行，然后用两手同时相对转动两个脚螺旋，直到气泡居中，注意气泡移动方向始终与左手大拇指移动方向一致。再将照准部旋转 90° ，如图 3-4（b）所示，转动第三个脚螺旋，使水管气泡居中，按以上步骤反复进行操作，直到照准部转至任意位置气泡皆居中为止。

Secondly, accurate leveling. As shown in Figure 3-4(a), when leveling, first rotate the alidade of the instrument to make the alidade level tube parallel to the connecting line of any pair of foot screws, and then rotate the two leg screws with both hands at the same time until the bubble is centered. Note that the moving direction of the bubble should be always consistent with the moving direction of the left thumb. Rotate the alidade by 90°, as shown in Figure 3-4(b), and turn the third foot screw to center the level tube bubble. Repeat the above steps until the bubble is always centered when the alidade is rotated to any position.

在精确整平后，需检查仪器对中情况。若测站点标志不在照准圈中心且偏移量较小，可松开仪器中心连接螺旋，在架头上平移仪器使其精确对中。在平移仪器时，整平可能受到影响，

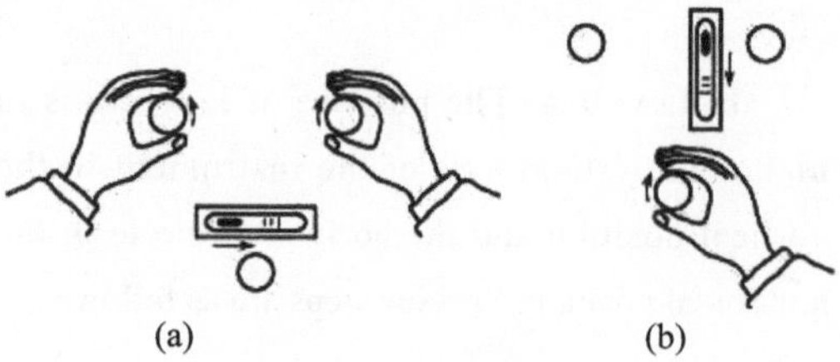

图3-4　整平图

Figure 3-4 Leveling

因此需要再精确整平。在精确整平时，对中又可能受到影响，于是，这两项工作需要反复进行，直到两者都满足为止。

After accurate leveling, check the instrument centering. If the observation station mark is not at the center of the sighting circle and the deviation is small, loosen the connecting screw at the center of the instrument and translate the instrument on the tripod head to make it accurately centered. Since leveling may be affected when the instrument is translated, accurate leveling is required. The accurate leveling may also affect the centering, so centering is required again. Therefore, these two tasks need to be repeated until both requirements are met.

②瞄准目标。测角时的照准标志，一般是竖立于测点的标杆、测钎及用三根竹竿悬吊垂球的线或觇牌。测量水平角时，以望远镜的十字丝竖丝瞄准照准标志。望远镜瞄准目标的操作步骤如下。

② Aiming at the target. The sighting marks for angle measurement are generally the ranging poles and marking pins erected at the measuring points, as well as the lines or targets suspended by three bamboo poles. When measuring the horizontal angle, aim the vertical wire of the telescope crosshair at the sighting mark. The operation steps are as follows:

a. 目镜对光。松开望远镜制动螺旋和水平制动螺旋，将望远镜对向明亮的背景（如白墙、天空等，注意不要对向太阳），转动目镜使十字丝清晰可见。

a. Eyepiece focusing. Loosen the telescope clamp screw and horizontal clamp screw, rotate the telescope towards a bright background (such as white wall and sky, but not the sun), and turn the eyepiece to make the crosshair clear to be seen.

b. 粗瞄目标。利用粗瞄器粗略瞄准目标后，旋紧望远镜制动螺旋和照准部制动螺旋。

b. Coarse sighting. After roughly aiming at the target with the coarse sighting device, tighten the telescope clamp screw and the alidade clamp screw.

c. 物镜对光。转动物镜对光螺旋，使目标影像清晰，并消除视差。

c. Objective lens focusing. Rotate the focusing screw of the objective lens to make the target image clear and eliminate parallax.

d. 精确瞄准。用望远镜微动螺旋和水平微动螺旋精确瞄准目标，瞄准目标时，应尽量瞄准目标底部，使用竖丝的中间部分平分或夹准目标。如图 3-5 所示。

d. Accurate sighting. Use the telescope fine-motion screw and horizontal fine-motion screw to accurately aim at the target. When aiming at the target, try to aim at the bottom of the target, and use the middle part of the vertical wire of the crosshair to bisect or clamp the target, as shown in Figure 3-5.

③读数。

③ Reading.

分微尺测微器及其读数方法。$DJ_6$ 型光学

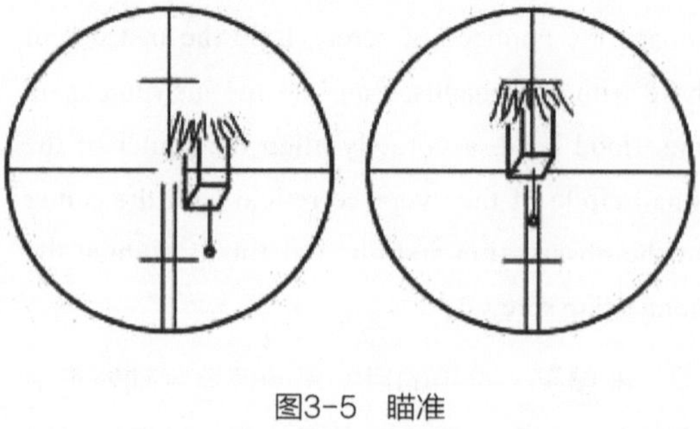

图3-5　瞄准

Figure 3-5 Sighting

经纬仪采用分微尺测微器进行读数。这类仪器的度盘分划值为 1°，按顺时针方向注记每度的度数。在读数显微镜的读数窗上装有一块带分划的分微尺，度盘上的分划线间隔经显微镜物镜放大后成像于分微尺上。图 3-6 所示读数为显微镜内所看到的度盘和分微尺的影像，上面注有“H”(或“水平”)的为水平度盘读数窗，注有“V”(或“竖直”)的为竖直度盘读数窗，分微尺的长度等于放大后度盘分划线间隔 1° 的长度，分微尺分为 60 个小格，每小格为 1′。分微尺每 10 小格注有数字，表示 0′、10′、20′、……、60′，注记增加方向与度盘相反。读数装置直接读到 1′，估读到 0.1′ (6″)。

Micrometer and its reading method. $DJ_6$ optical theodolite uses micrometer for reading. Its graduation value is 1°, and the degrees are marked clockwise. There is a micrometer with graduations installed on the reading window of the reading microscope, and graduations on the circle are imaged on the micrometer after magnified by the microscope objective lens. The readings shown in Figure 3-6 are the images of the circle and micrometer seen by the microscope, with “H” (or “horizontal”) standing for horizontal circle and “V” (or “vertical”) standing for the vertical circle. The length of the micrometer is equal to the magnified length of a 1° interval of the circle. The micrometer is divided into 60 scales, with each scale being 1′ and every 10 scales marked with figures, namely 0′, 10′, 20′, ..., 60′, which is in an inverse order of the increasing marking direction of the circle. The reading device reads directly to 1′ and estimates to 0.1′ (6″).

读数时，分微尺上的 0 分划线为指标线，它在度盘上的位置就是度盘读数的位置。如在水平度盘的读数窗中，分微尺的 0 分划线已超过 261°，水平度盘的读数应该多于 261°。所多的数值，再由分微尺的 0 分划线至度盘上 261 分划线之间有多少小格来确定。图 3-6 中

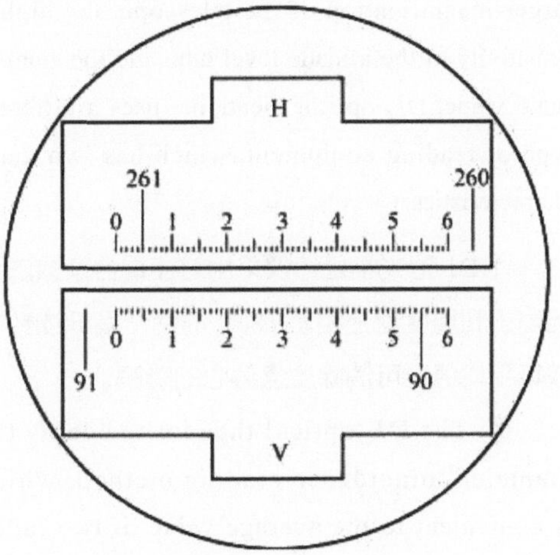


图3-6　分微尺测微器

Figure 3-6 Micrometer

为 4.4 格，故为 04′ 24″。水平度盘的读数应是 261° 04′ 24″。

For reading, the 0 division line on the micrometer is the index line, and its position on the circle is the position of the circle reading. If the 0 division line of the micrometer is beyond 261° in the reading window of the horizontal circle, the reading of the horizontal circle should be more than 261°. The extra value is determined by the number of divisions between the 0 division of the micrometer and the 261° division line of the circle. As shown in Figure 3-6, there are 4.4 scales, so it is 04′ 24″. The reading of the horizontal circle shall be 261°04′ 24″.

## (2) $DJ_2$ 型光学经纬仪的读数方法

### (2) Reading method of $DJ_2$ optical theodolite

$DJ_2$ 型光学经纬仪的观测精度高于 $DJ_6$ 型光学经纬仪。在结构上，除望远镜的放大倍数较大、照准部水准管的灵敏度较高、度盘格值较小外，主要表现为读数设备的不同。$DJ_2$ 型光学经纬仪的读数设备主要有两个特点。

The observation accuracy of $DJ_2$ optical theodolite is higher than that of $DJ_6$ optical theodolite. In terms of the structure, except for the larger magnification of the telescope, the higher sensitivity of the alidade level tube and the smaller scale value, $DJ_2$ optical theodolite uses a different type of reading equipment, which has two main characteristics.

①$DJ_2$型光学经纬仪采用对径重合读数法，相当于利用度盘上相差 180° 的两个指标读数并取其平均值，可消除度盘偏心的影响。

① The $DJ_2$ optical theodolite adopts the diameter coincidence reading method, which is equivalent to the average value of two index readings with a difference of 180° on the circle and can eliminate the influence of the eccentricity of the circle.

② $DJ_2$ 型光学经纬仪在读数显微镜中只能看到水平度盘或竖直度盘中的一种，读数时，可通过转动度盘变换手轮，选择所需要的度盘影像。

② Only either horizontal circle or vertical circle is displayed in the reading microscope of the $DJ_2$ optical theodolite. When reading, rotate the image changing handwheel to select the required circle images.

$DJ_2$ 型光学经纬仪利用度盘 180° 对径分划线影像符合读数装置进行读数。外部光线进入仪器后，经过一系列棱镜和透镜的作用，将度盘上直径两端分划，同时，反映到读数显微镜的中间窗口，呈方格状。当转动测微轮时，呈上、下两部分的对径分划的影像将做相对移动，当上下分划的影像精准重合时才能读数。

$DJ_2$ optical theodolite uses the image coincidence reading device for reading after images above and below the 180° diameter division line of the circle coincide. After the external light enters the instrument, the two ends of the diameter on the circle are divided by a series of prisms and lenses, and at the same time, it is reflected in the middle window of the reading microscope in square lattices. When the micrometer wheel is rotated, the images above and below the 180° diameter division line will move relatively, and the reading can only be taken when the images accurately coincide.

近年生产的 $DJ_2$ 型光学经纬仪采用了数字化读数装置，读数窗中用数字显示整 10′ 数，如图 3-7(a) 所示。读数时，先转动测微轮，使度盘的主、副像分划线重合，如图 3-7(b) 所示，然后读数，65° 54′ 03″ 。

The $DJ_2$ optical theodolite produced in recent years is equipped with a digital reading device, and the reading window displays in an integral multiple of 10′ , as shown in Figure 3-7(a). When reading, rotate the micrometer wheel to make the main and auxiliary image division lines of the circle coincide, as shown in Figure 3-7(b), and then read out, that is 65°54′ 03″ .

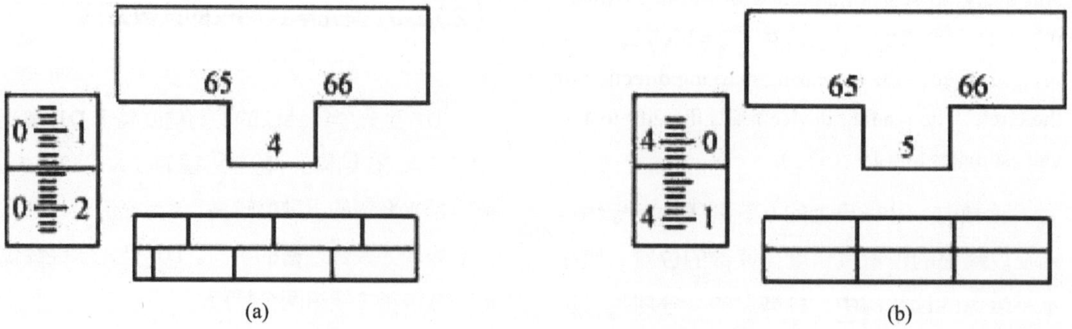


(a)　　(b)

图3-7　$DJ_2$型光学经纬仪度盘读数

Figure 3-7 Circle Reading of $DJ_2$ Optical Theodolite

# 3.3 水平角测量的方法

## 3.3 Measurement Methods of Horizontal Angle

普通测量中常用的水平角观测方法有测回法和方向观测法两种。

The horizontal angle observation methods commonly used in general survey include observation set method and direction observation method.

无论采用哪种方法进行水平角观测，通常都要用盘左和盘右各观测一次。所谓盘左，就是竖盘位于望远镜的左边，又称为正镜；盘右就是竖盘位于望远镜的右边，又称为倒镜。将正、倒镜的观测结果取平均值，可以抵消部分仪器误差的影响，提高测量结果准确度。如果只用盘左或者盘右观测一次，称为半个测回或半测回；如果用盘左、盘右各观测一次，称为一个测回或一测回。

Regardless of the method used for horizontal angle observation, it is usually required that observation shall be carried out once by using the face left and face right positions, respectively. Face left position means that the vertical circle is located on the left side of the telescope, also known as direct telescope; face right position means that the vertical circle is located on the right side of the telescope, also known as telescope in reversed telescope. Taking the average value of the observation results measured by placing the telescope in face left position and face right position can offset the influence of some instrument errors and improve the accuracy of the measurement results. If only one observation is carried out by using either face left position or face right position, it is called half observation set or semi-observation; if observation is carried out once by using face left position and face right position, it is called an observation set or one observation set.

下面介绍用测回法和方向观测法测量水平角的操作步骤和记录、计算方法。

The operation steps as well as the recording and calculation methods of observation set method and direction observation method are described below.

### 3.3.1 测回法

### 3.3.1 Observation Set Method

当所测的角度只有两个方向时，通常都用测回法测量。如图 3-8 所示，$A$、$B$ 为观测目标，用测回法观测 $OA$ 与 $OB$ 两方向之间的水平角，具体施测步骤如下。

When the angle to be measured has only two directions, the observation set method is usually used for observation. As shown in Figure 3-8, $A$ and $B$ are the observed targets, and the horizontal angle between $OA$ and $OB$ is observed by the observation set method. The specific measurement steps are as follows.

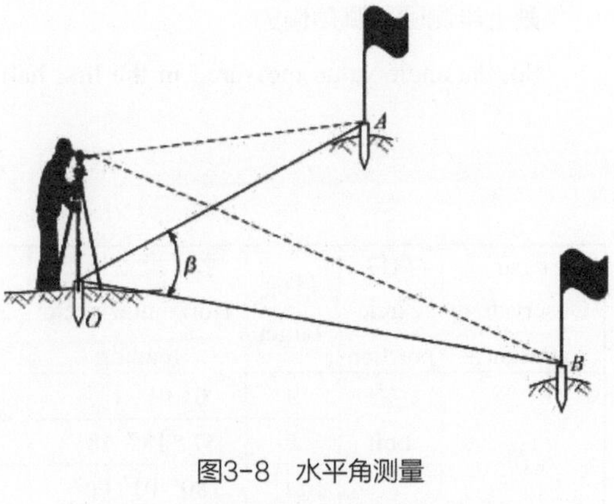


图3-8　水平角测量

Figure 3-8 Measurement of Horizontal Angle

①将仪器安置在 $O$ 点上，对中、整平。

① Place the instrument at point $O$, center and level it.

②盘左位置，松开水平制动螺旋和望远镜制动螺旋，用望远镜上的准星、照门或粗瞄器瞄准左目标 $A$，旋紧两制动螺旋，进行目镜和物镜对光，使十字丝和目标成像清晰，消除视差。再用水平微动螺旋和望远镜微动螺旋精确瞄准目标的下部，读取水平度盘读数记入观测手簿（表 3-1）。松开水平制动螺旋，转动照准部，以同样的方法瞄准右目标 $B$，读取水平度盘读数为 57° 18′ 48″，记入手簿。

② In the face left position, release the horizontal clamp screw and the telescope clamp screw, aim at the left target $A$ with the front sight, rear sight or coarse sighting device on the telescope, tighten the two clamp screws for focusing of the eyepiece and objective lens to make the crosshair and target image clear and eliminate parallax. Then, use the horizontal fine-motion screw and the telescope fine-motion screw to accurately aim at the lower part of the target, read the horizontal circle reading, and record it in the record book (Table 3-1). Loosen the horizontal clamp screw, turn the alidade, aim at the right target $B$ in the same way, read the horizontal circle reading as 57°18′48″, and record it in the record book.

则上半测回所测角值为

So, the angle value measured in the first half observation set is

$$\beta_1=b_1-a_1=57°18'48''-0°01'12''=57°17'36''$$

③盘右位置，松开照准部制动螺旋，倒镜成盘右位置，先瞄准右目标 $B$，读取水平度盘读数为 237° 18′ 54″，记入观测手簿。松开照准部制动螺旋，逆时针转动照准部，瞄准左目标 $A$，读取水平度盘读数为 180° 01′ 06″，记入手簿。

③ For the face right position, release the alidade clamp screw, reverse the telescope to the face right position, aim at the right target $B$ first, read the horizontal circle reading as 237°18′54″, and record it in the record book. Loosen the alidade clamp screw, turn the alidade counterclockwise, aim at the left target $A$, read the horizontal circle reading as 180°01′06″, and record it in the record book.

下半测回所测角值为

So, the angle value measured in the second half observation set is

$$\beta_2=b_2-a_2$$

$$=237°18'54''-180°01'06''$$

$$=57°17'48''$$

上半测回和下半测回构成一测回。

The first half observation set and the second half observation set constitute one observation set.

表 3-1　测回法观测手簿

Table 3-1 Record Book for Observation Set Method

| 测站 Observation station | 盘位 Circle position | 目标 Target | 水平度盘读数 Horizontal circle reading | 半测回角值 Angle measured in half observation set | 一测回角值 Angle measured in one observation set | 备注 Remarks |
|---|---|---|---|---|---|---|
| $O$ | 左 Left | $A$ | 0°01′ 12″ | 57° 17′ 36″ | 57° 17′ 42″ | |
| | | $B$ | 57°18′ 48″ | | | |
| | 右 Right | $A$ | 180° 01′ 06″ | 57° 17′ 48″ | | |
| | | $B$ | 237° 18′ 54″ | | | |

④对于 $DJ_6$ 型光学经纬仪，如果上、下两半测回角值之差不大于 ±40″，则认为观测合格。此时，取上、下两半测回角值的平均值作为一测回角值。

④ For the $DJ_6$ optical theodolite, if the difference between angle values measured in the first and second half observation sets is not more than ±40″, it is considered that the observation is qualified. At this time, the average value of the angle values measured in the first and second half observation sets is taken as the angle value of one observation set.

## 3.3.2 方向观测法

### 3.3.2 Direction Observation Method

当在一个测站上需观测多个方向时，宜采用方向观测法（又称全圆测回法），因为可以简化外业工作。它的直接观测结果是各个方向相对于起始方向的水平角值，也称为方向值。相邻方向的方向值之差，就是它的水平角值。如图 3-9 所示，*O* 为测站点，*A*、*B*、*C*、*D* 为观测目标，用方向观测法观测各方向的水平角，其观测、记录和计算方法如下。

When multiple directions need to be observed at an observation station, the direction observation method (also known as the full circle observation set method) should be adopted for simplified field work. Its direct observation result is the horizontal angle value of each direction relative to the starting direction, also called the direction value. The difference between the direction values of adjacent directions is the horizontal angle value. As shown in Figure 3-9, *O* is the observation station, and *A*, *B*, *C* and *D* are the observation targets. The horizontal angles between various direction are observed by the direction observation method. The horizontal angle observation, recording and calculation methods are as follows.

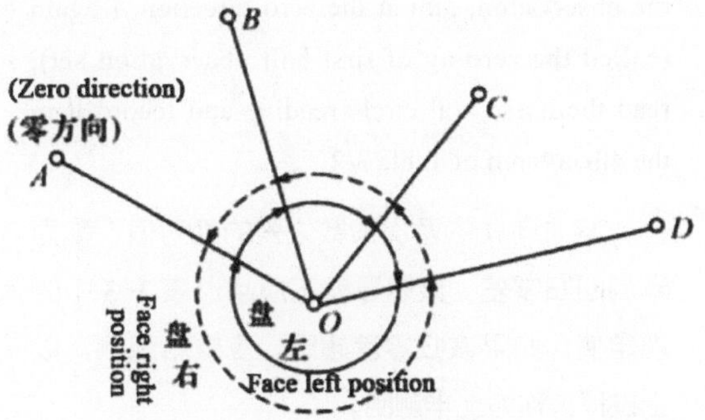


图3-9　方向观测法

Figure 3-9 Direction Observation Method

①在测站点 *O* 安置经纬仪，在 *A*、*B*、*C*、*D* 观测目标处竖立观测标志。

① Place the theodolite at observation station *O*, and erect observation marks at observation targets *A*, *B*, *C* and *D*.

②盘左位置，选择方向中一个明显目标（如 *A*）作为起始方向（或称零方向），精确瞄准 *A*，将水平度盘配置为略大于 0，将水平度盘读取读数记入表 3-2 所示方向观测法观测手簿的第 4 列。

② For the face left position, select an obvious target, for example *A*, as the starting direction (or zero direction), accurately aim at *A*, set the horizontal circle to be slightly greater than 0°, and record the reading in the 4th column of the record book for direction observation method.

松开照准部制动螺旋，顺时针方向旋转照准部，依次瞄准 *B*、*C*、*D* 各目标，分别读取水平度盘读数，记入表 3-2 中的第 4 列。为了检核水平度盘在观测过程中是否发生变动，应再次瞄准零方向 *A*（称为上半测回归零），读取水平度盘读数，记入表 3-2 中的第 4 列。

Loosen the alidade clamp screw, rotate the alidade clockwise, aim at targets *B*, *C* and *D* in turn, read the horizontal circle readings respectively and record them in column 4 of Table 3-2. In order to check whether the horizontal circle changes during the observation, aim at the zero direction $A$ again (called the zeroing of first half observation set), read the horizontal circle reading and record it in the 4th column of Table 3-2.

零方向 $A$ 的两次读数之差的绝对值，称为半测回归零差，该归零差不应超过表 3-3 中的规定值；如果该归零差超限，应重新观测。以上过程被称为上半测回。

The absolute value of the difference between the two readings in the zero direction $A$ is called the semi-observation misclosure of round, which shall not exceed the value specified in Table 3-3. If the misclosure of round exceeds the specified value, carry out the observation again. The above process is called the first half observation set.

③盘右位置，按逆时针方向依次照准目标 $A$、$D$、$C$、$B$、$A$，并将水平度盘读数由下向上记入表 3- 2 中的第 5 列，此过程被为下半测回。

③ For the face right position, aim at the targets $A$, $D$, $C$, $B$ and $A$ counterclockwise in turn, and record the horizontal circle readings in the 5th column of Table 3-2 from bottom to top. This process is called the second half observation set.

上、下两个半测回合称一测回。为了提高精度，有时需要观测 $n$ 个测回，各测回起始方向仍按 180° /$n$ 的差值安置水平度盘读数。

The first half observation set and the second half observation set together are called one observation set. In order to improve the accuracy, it is sometimes necessary to observe $n$ observation sets, and the horizontal circle reading is still configured according to the difference of 180°/$n$ for the starting direction of each observation set.

表 3-2 方向观测法观测手簿

Table 3-2 Record Book for Direction Observation Method

| 测站 Observation station | 测回数 Number of observation sets | 目标 Target | 水平度盘读数 Horizontal circle reading | | 2$c$ | 平均读数 Average reading | 归零后方向值 Direction value after zeroing | 各测回归零后方向平均值 Average direction value of various observation sets after zeroing | 角值 Angle value |
|---|---|---|---|---|---|---|---|---|---|
| | | | 盘左 Face left position | 盘右 Face right position | | | | | |
| 1 | 2 | 3 | 4 | 5 | 6 | 7 | 8 | 9 | 10 |
| O | 1 | $A$ | 0° 02′ 10″ | 180° 02′ 02″ | + 8″ | (0° 02′ 08″ ) 0° 02′ 06 | 0° 00′ 00″ | 0° 00′ 00″ | |
| | | $B$ | 37° 44′ 10″ | 217° 44′ 05″ | + 5″ | 37° 44′ 08″ | 37° 42′ 00″ | 37° 42′ 03″ | 37° 42′ 03″ |
| | | $C$ | 110° 29′ 02″ | 290° 28′ 54″ | + 8″ | 110° 28′ 58″ | 110° 26′ 50″ | 110° 26′ 54″ | 72° 44′ 51″ |
| | | $D$ | 150° 14′ 52″ | 330° 14′ 46″ | + 6″ | 150° 14′ 49″ | 150° 12′ 41″ | 150° 12′ 34″ | 39° 45′ 40″ |
| | | $A$ | 0° 02′ 12″ | 180° 02′ 08″ | + 4″ | 0° 02′ 10″ | | | |

续表

| 测站 Observation station | 测回数 Number of observation sets | 目标 Target | 水平度盘读数 Horizontal circle reading | | 2c | 平均读数 Average reading | 归零后方向值 Direction value after zeroing | 各测回归零后方向平均值 Average direction value of various observation sets after zeroing | 角值 Angle value |
|---|---|---|---|---|---|---|---|---|---|
| | | | 盘左 Face left position | 盘右 Face right position | | | | | |
| *O* | 2 | *A* | 90°03′30″ | 270°03′26″ | +4″ | (90°03′26″) 90°03′28″ | 0°00′00″ | | |
| | | *B* | 127°45′36″ | 307°45′28″ | +8″ | 127°45′32″ | 37°42′06″ | | |
| | | *C* | 200°30′26″ | 20°30′20″ | +6″ | 200°30′23″ | 110°26′57″ | | |
| | | *D* | 240°15′58″ | 60°15′50″ | +8″ | 240°15′54″ | 150°12′28″ | | |
| | | *A* | 90°03′26″ | 270°03′20″ | +6″ | 90°03′23″ | | | |

④计算方法。

④ Calculation method.

a. 计算两倍视准轴误差 $2c$ 值（一个测回内盘左与盘右的差值）。

a.Calculate the $2c$ value, i.e., difference of the discrepancy between twice collimation errors.

$$2c = \text{盘左读数} - (\text{盘右读数} \pm 180°) \quad (3.2)$$

$$2c = \text{Face left position reading} - (\text{Face right position reading} \pm 180°) \quad (3.2)$$

上式中，盘右读数大于 180° 时取“−”号，盘右读数小于 180° 时取“+”号。计算各方向的 2c 值，填入表 3-2 中的第 6 列。一测回内各方向 2c 值互差不应超过表 3-3 中的规定值。如果超限，应在原度盘位置重测。

When the face right position reading is larger than 180°, take the “minus” value; when the face right position reading is smaller than 180°, take the “plus” value. Calculate the 2c values of various directions and fill them in the 6th column of Table 3-2. The difference between every 2c values of various directions measured in the first observation set shall not exceed the value specified in Table 3-3. If it exceeds the specified value, observation shall be carried out again in the original circle position.

b. 计算各方向的平均读数。平均读数为各方向的平均方向值。

b.Calculate the average reading of various directions. The average reading is the average direction value of various directions.

$$\text{平均读数} = \frac{1}{2}[\text{盘左读数} + (\text{盘右读数} \pm 180°)] \quad (3.3)$$

$$\text{Average reading} = \frac{1}{2}[\text{Face left position reading} + (\text{Face right position reading} \pm 180°)] \quad (3.3)$$

计算时，以盘左读数为准，将盘右读数加或减 180°，再和盘左读数相加后取平均值。计算各方向的平均读数，填入表 3-2 中的第 7 列。起始方向有两个平均读数，故应再取其平均值，填入表 3-2 中第 7 列的括号内。

For calculation, first add or subtract the face right position reading by 180°, add the face left position reading, and average the sum. Calculate the average value of various directions and fill it in the 7th column of Table 3-2. There are two average readings in the starting direction, so take the average value again and fill it in the bracket in the 7th column of Table 3-2.

c. 计算归零后的方向值。将各方向的平均读数减去起始方向的平均读数（括号内数值），即得各方向的“归零后方向值”，填入表 3-2 中的第 8 列。起始方向归零后的方向值为零。

c. Calculate the direction value after zeroing. Subtract the average reading of each direction by the average reading of the starting direction (the value in bracket) to obtain the “direction value in each direction after zeroing”, and fill it in the 8th column of Table 3-2. The direction value of the starting direction after zeroing is zero.

d. 计算各测回归零后方向值的平均值。多测回观测时，同一方向值各测回互差符合表 3-3 中的规定，则取各测回归零后方向值的平均值，作为该方向的最后结果，填入表 3-2 中的第 9 列。

d.Calculate the average direction value of the direction values measured in various observation sets after zeroing. In case of multiple observation sets, if difference between the direction values measured in various observation sets for the same direction is within the value specified in Table 3-3, the average value of the direction values measured in various observation sets after zeroing is taken as the final result for this direction and filled in the 9th column of Table 3-2.

e. 计算各目标间水平角角值。将表 3-2 中第 9 列相邻两方向值相减即可求得，填入表 3-2 中第 10 列相应位置上。

e.Calculate the horizontal angle between targets Subtract the values in the adjacent two direction values in the 9th column of Table 3-2, and record the difference values in the 10th column of Table 3-2.

表 3-3　限差规定值

Table 3-3 Limited Difference Specified Value

| 仪器 Instru-ment | 半测回归零差/(″) Semi-observation misclosure of round/(") | 一测回内 $2c$ 互差/(″) $2c$ difference within one observation set/(") | 同一方向值各测回互差/(″) Difference between various observation sets for the same direction/(") | 仪器 Instru-ment | 半测回归零差/(″) Semi-observation misclosure of round/(") | 一测回内 $2c$ 互差/(″) $2c$ difference within one observation set/(") | 同一方向值各测回互差/(″) Difference between various observation sets for the same direction/(") |
|---|---|---|---|---|---|---|---|
| $J_2$ | 12 | 18 | 12 | | 18 | | 24 |

当需要观测的方向为三个时，除不做归零观测外，其他均与三个以上方向的观测方法相同。

When observation is required for three directions, the observation method is the same as that for more than three directions, except that the repeated observation after zeroing is not required.

# 3.4 竖直角测量的方法

# 3. 4 Measurement Methods of Vertical Angle

## 3.4.1 竖直度盘构造

## 3.4.1 Structure of Vertical Circle

光学经纬仪竖直度盘的构造包括竖盘、竖盘指标、竖盘指标水准管和竖盘指标水准管微动螺旋。如图 3-10 所示，竖盘固定在望远镜横轴的一端，其面与横轴垂直。望远镜横轴旋转时，竖盘也随之旋转，而竖盘指标不动。竖盘指标为分微尺的零分划线，它与竖盘指标水准管固连在一起，当旋转竖盘指标水准管微动螺旋使指标水准管气泡居中时，竖盘指标即处于正确位置。有些经纬仪采用了竖盘指标自动归零装置，打开自动补偿器，竖盘指标即居于正确位置，提高了竖直角观测的速度和精度。

An optical theodolite consists of a vertical circle, vertical circle index, vertical circle index level tube and vertical circle index level tube micro-motion screw. As shown in Figure 3-10, the vertical circle is fixed at one end of the horizontal axis of the telescope, and its surface is perpendicular to the horizontal axis. When the horizontal axis of the telescope rotates, the vertical circle also rotates, while the vertical circle index does not move. The vertical circle index is the zero-division line of the micrometer, which is fixedly connected with the vertical circle index level tube. When the vertical circle index level tube micro-motion screw is rotated to make the level tube bubble centered, the vertical circle index is in the correct position. Some theodolites adopt the vertical circle index automatic zeroing device, and the vertical circle index is in the correct position when the automatic compensator is turned on, which improves the speed and accuracy of vertical angle observation and measurement.

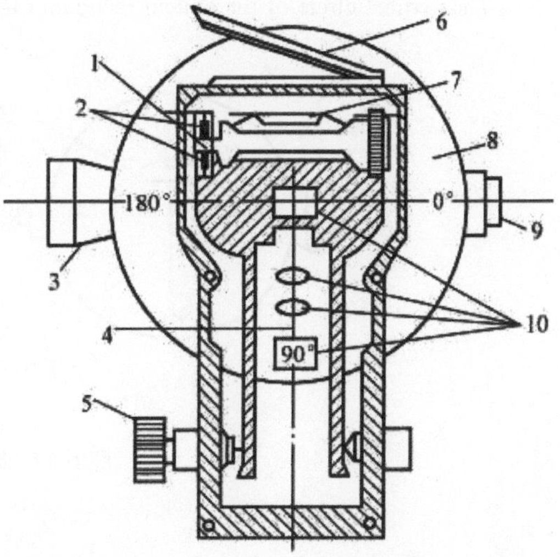


1—指标水准管；2—标水准管校正螺丝；3—望远镜；4—光具组光轴；5—光指标水准管微动螺旋；6—标指标水准管反光镜；7—标指标水准管；8—竖盘；9—竖目镜；10—光具组的透镜和棱镜。

1-index level tube; 2-level tube correction screw; 3-telescope; 4-optical axis of optical system; 5-index level tube micro-motion screw; 6-index level tube reflector; 7-index level tube; 8-vertical circle; 9-eyepiece; 10-lens and prisms of optical system.

图3-10 竖直度盘的构造

Figure 3-10 Structure of Vertical Circle

竖盘的注记形式有顺时针与逆时针两种。当望远镜视线水平，竖盘指标水准管气泡居中时，盘左竖盘读数应为 90° ，盘右竖盘读数应为 270° 。

There are two types of scale marking for the vertical circle: clockwise and counterclockwise. When the line of sight of the telescope is horizontal and the bubble of the vertical circle index level tube is centered, the reading of the vertical circle in the face left position shall be 90°, and the reading of the vertical circle in the face right position shall be 270°.

光学经纬仪的竖直度盘也是一个玻璃圆环，分划与水平度盘相似，度盘刻度 0° ~360° 的注记，有顺时针方向和逆时针方向两种。图 3-11(a) 所示为顺时针方向注记，图 3-11(b) 所示为逆时针方向注记。

The vertical circle of the optical theodolite is also a glass ring, and the graduation is similar to the horizontal dial. The scale of the vertical circle is marked either clockwise or counterclockwise from 0° to 360°. Figure 3-11(a) shows the clockwise scale marking; Figure 3-11 (b) shows the counterclockwise scale marking.

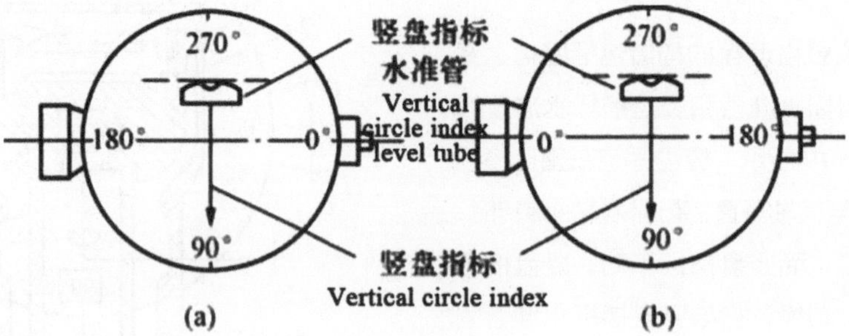


图3-11 竖直度盘刻度注记

Figure 3-11 Scale Marking of Vertical Circle

## 3.4.2 竖直角计算公式

### 3.4.2 Calculation Formula of Vertical Angle

由于竖盘注记形式不同，竖直角计算的公式也不一样。现在以顺时针注记的竖盘为例，推导竖直角的计算公式。

Due to the different types of vertical circle scale marking, the calculation formula of the vertical angle is also different. Now take the vertical circle with clockwise scale marking as an example to derive the calculation formula of the vertical angle.

盘左位置如图 3-12(a) 所示：视线水平时，竖盘读数为 90° 。当瞄准一目标时，竖盘读数为 $L$，则盘左竖直角为

As shown in Figure 3-12 (a), in the face left position, when the line of sight is horizontal, the reading of the vertical circle is 90°. When a target is sighted, the vertical circle reading is $L$, and the face left vertical angle is

$$\alpha_L = 90° - L \tag{3.4}$$

盘右位置如图 3-12(b) 所示：视线水平时，竖盘读数为 270° 。当瞄准同一目标时，竖盘读数为 $R$，则盘右竖直角为

As shown in Figure 3-12 (b), in the face right position, when the line of sight is horizontal, the reading of the vertical circle is 270°. When the same original target is sighted, the vertical circle reading is $R$, and the vertical angle in the face right position is

$$\alpha_R = R - 270° \tag{3.5}$$

将盘左、盘右位置的两个竖直角取平均值，即得竖直角 $\alpha$ 的计算公式为

Taking the average value of the two vertical angles in the face left and right positions to obtain the vertical angle $\alpha$, the following formula is expressed as

$$\alpha = \frac{1}{2}(\alpha_L + \alpha_R) \tag{3.6}$$

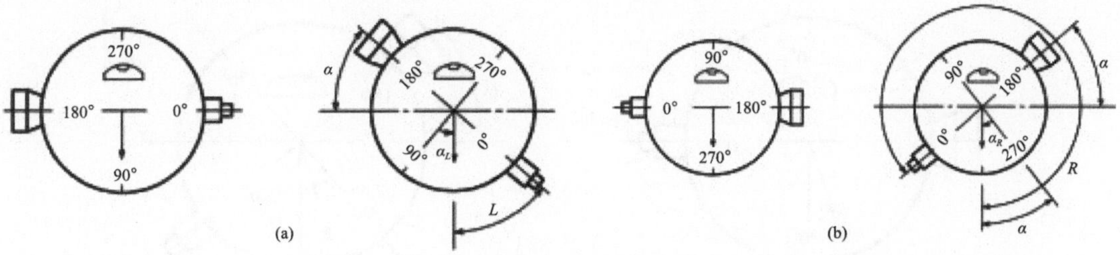


图3-12　竖盘注记

Figure 3-12  Vertical Circle Scale Marking

### 3.4.3 竖盘指标差

### 3.4.3 Index Error of Vertical Circle

在竖直角计算公式中，当视准轴水平、竖盘指标水准管气泡居中时，竖盘读数应是 90° 的整数倍。但是实际上这个条件往往不能满足，竖盘指标常常偏离正确位置，这个偏离的差值称为竖盘指标差，用 $x$ 表示。竖盘指标差 $x$ 本身有正负号，一般规定当竖盘指标偏移方向与竖盘注记方向一致时，$x$ 取正号；反之，$x$ 取负号。

In the vertical angle calculation formula, when the collimation axis is horizontal and the vertical circle index level tube bubble is centered, the vertical circle reading shall be an integral multiple of 90°. However, in fact, this condition cannot be always met, and the vertical circle index often deviates from the correct position. The difference of this deviation is called the index error of vertical circle, which is expressed by $x$. The index error of vertical circle $x$ itself has plus and minus signs. Generally, when the direction of the vertical circle index deviation is consistent with the vertical circle scale marking direction, take the plus sign for $x$ or take the minus sign.

盘左位置 [ 图 3-13(a)] 由于存在指标差，其正确的竖直角计算公式为

For the face left position as shown in Figure 3-13(a), due to the index error, the correct vertical angle calculation formula is

$$\alpha_L = 90° - L + x \quad (3.7)$$

而盘右位置 [ 图 3-13(b)]，其正确的竖直角计算公式为

For the face right position, as shown in Figure 3-13(b), the correct vertical angle calculation formula is

$$\alpha_R = R - 270° - x \quad (3.8)$$

将式（3.7）和式（3.8）相加并除以 2，得

Add formula (3.7) and formula (3.8) and divide the sum by 2 to obtain the following formula

$$\alpha = \frac{1}{2}(\alpha_L + \alpha_R) = \frac{1}{2}(R - L - 180°) \quad (3.9)$$

由此可见，在竖直角测量时，用盘左、盘右观测，取平均值作为竖直角的观测结果，可以消除竖盘指标差的影响。

Therefore, when measuring the vertical angle, one can take the face left and right positions for observation and take the average value of the two vertical angle values measured, so as to eliminate the influence of the vertical circle index error.

将式（3.7）和式（3.8）相减并除以 2，得

Subtract formula (3.7) and formula (3,8) and divide the difference by 2 to obtain the following formula

$$x = \frac{1}{2}(\alpha_L - \alpha_R + L + R - 360°) \quad (3.10)$$

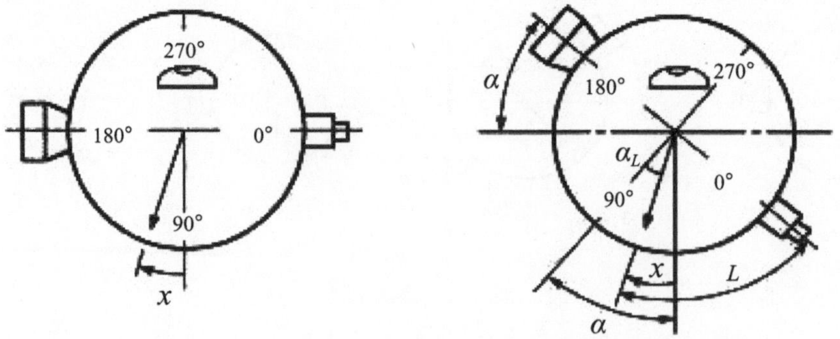


(a)

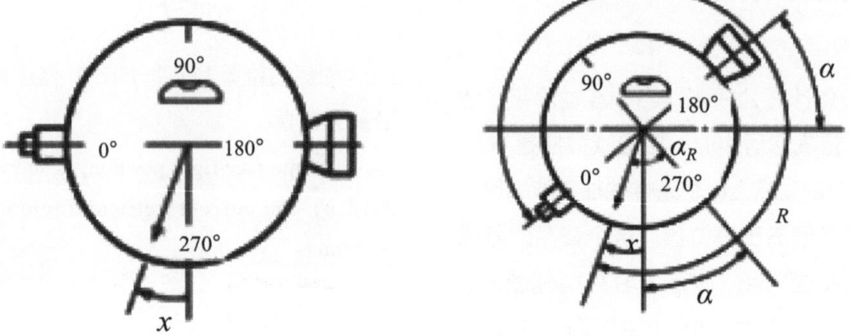


(b)

图3-13　竖盘指标差

Figure 3-13 Index Error of Vertical Circle

式 (3.10) 为竖盘指标差的计算公式。指标差互差（即所求指标差之间的差值）可以反映观测结果的精度。有关规范规定：竖直角观测时，指标差互差的限差，$DJ_2$ 型仪器不得超过 ±15″；$DJ_6$ 型仪器不得超过 ±25″。

Formula (3.10) is the calculation formula of vertical circle index error. Difference between the index errors can reflect the accuracy of observation results. Relevant specifications stipulate that the tolerance of deference between index errors shall not exceed ±15″for $DJ_2$ instrument and ±25″for $DJ_6$ instrument during vertical angle observation.

## 3.4.4 竖直角观测

### 3.4.4 Observation of Vertical Angle

由竖直角的定义可知，它是倾斜视线与在同一铅垂面内的水平视线所夹的角度。水平视线的读数为已知常数 (90° 或 270°)，故竖直角观测不必观测视线水平方向，只需观测目标点，并读取倾斜视线方向的竖盘读数，即可按前述公式算出竖直角值。但为了消除仪器误差的影响，同样需要用盘左、盘右观测。

From the definition of vertical angle, it can be seen that it is the angle between the oblique line of sight and the horizontal line of sight in the same vertical plane. Since the reading of the horizontal line of sight is a known constant (90° or 270°). Therefore, it is not necessary to observe the horizontal direction of the line of sight for vertical angle observation. Only the target point needs to be observed and the reading of the vertical circle in the inclined line of sight can be taken to calculate the vertical angle value according to the above formula.

However, in order to eliminate the influence of instrument error, it is also necessary to observe by taking both the face left and right positions.

用测回法（图 3-14）观测竖直角的测法、记录和计算步骤如下。

The measuring, recording and calculating methods for observing the vertical angle with the observation set method (Figure 3-14) are described as follows.

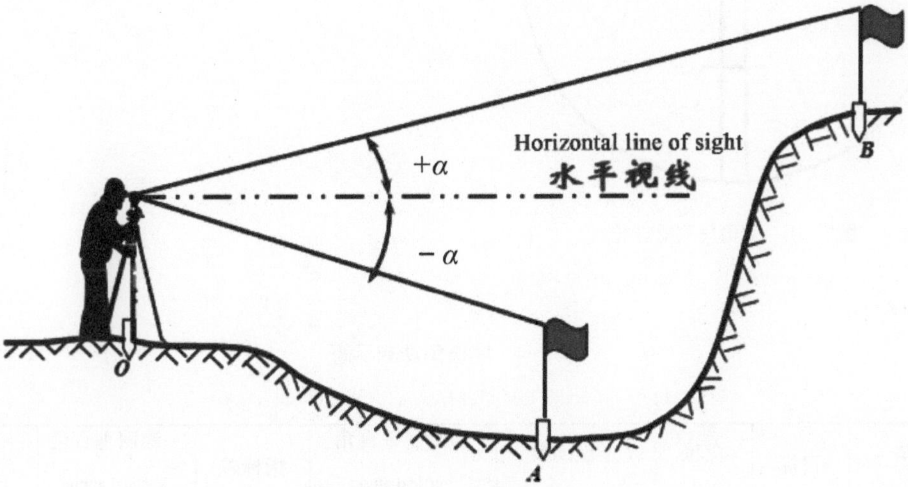


图3-14　观测竖直角

Figure 3-14 Observation of Vertical Angle

①在测站点 $O$ 安置经纬仪，在目标点 $A$ 竖立观测标志，确定该仪器竖直角的计算公式。为方便应用，可将公式记录于竖直角观测手簿（表 3-4）备注列中。

① Place the theodolite at the observation station $O$, erect the observation mark at the target point $A$, and determine the calculation formula of the vertical angle of the instrument. For the convenience of application, the formula can be recorded in the remarks column of Table 3-4 .

②盘左位置：瞄准目标 $A$，使十字丝横丝精确地切于目标顶端，如图 3-15 所示。转动竖盘指标水准管微动螺旋，使水准管气泡严格居中。然后，读取竖盘读数 $L$，设为 95° 22′ 00″，记入竖直角观测手簿（表 3-4）的相应位置。

② Circle left position: aim at target $A$ and make the crosshair accurately tangent to the top of the target, as shown in Figure 3-15. Rotate the vertical circle index level tube fine-motion screw to make the level tube bubble exactly centered. Then, read the vertical circle reading $L$, set it to 95°22′00″, and record it in the corresponding column of Table 3-4 .

③盘右位置：瞄准目标 $A$，使十字丝横丝精确地切于目标顶端，转动竖盘指标水准管微动螺旋，使水准管气泡严格居中。然后，读取竖盘读数 $R$，设为 264° 36′ 48″，记入表 3-4 的相应位置。

③ Circle right position: aim at target $A$, make the crosshair accurately tangent to the top of the target, and rotate the vertical circle index level tube fine-motion screw to make the level tube bubble exactly centered. Finally, take the vertical circle

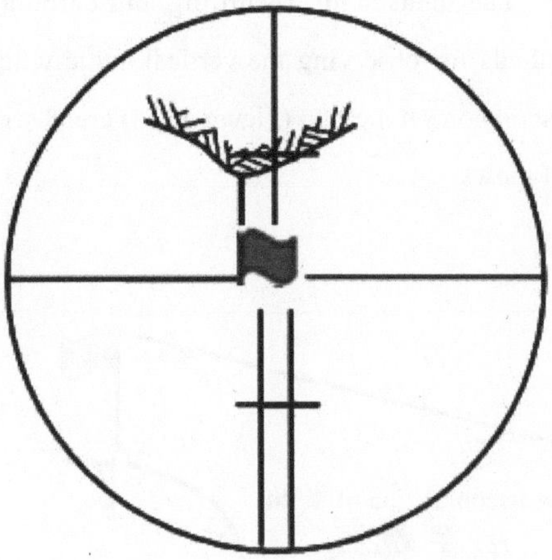

图3-15 竖直角测量瞄准

Figure 3-15 Vertical Angle Measurement Aim

reading $R$, set it to 264°36′ 48″, and record it in the corresponding column of Table 3-4.

表 3-4 竖直角观测手簿

Table 3-4 Record Book for Vertical Angle Observation

| 测站<br>Observation station | 目标<br>Target | 盘位<br>Circle position | 竖盘读数<br>Vertical circle reading | 半测回垂直角<br>Vertical angle measured in half observation set | 指标差<br>Index error | 一测回垂直角<br>Vertical angle measured in one observation set | 备注<br>Note |
|---|---|---|---|---|---|---|---|
| *O* | *A* | 左<br>Left | 95 °22′00″ | -5°22′00″ | -36″ | -5°22′36″ | |
| | | 右<br>Right | 264°36′48″ | -5°23′12″ | | | |

将计算结果分别填入表 3-4 的相应位置。

Fill the calculation results in the corresponding columns of Table 3-4.

# 本章小结

## Chapter Summary

本章主要讲述了水平角、竖直角测量原理，光学经纬仪的构造及使用，水平角与竖直角测量的方法、记录及计算、检验与校正，水平角测量误差产生的原因及注意事项等内容。

This chapter mainly describes the measurement principle of horizontal angle and vertical angle, the structure and use of optical theodolite, the measuring, recording and calculating method of horizontal angle and vertical angle, inspection and calibration of angle measurement, the causes and precautions of horizontal angle measurement error.

1. 水平角是地面上某一点到两个目标点的方向线垂直投影在水平面上的夹角，即通过这两条方向线所作两竖直面间的二面角，它是确定点的平面位置基本要素之一。竖直角（垂直角）是在同一竖直面内，某一点到目标的方向线与水平线之间的夹角，它是确定地面点高程位置的一个要素。

1.The horizontal angle is the included angle between the two direction lines from a certain point on the ground to two target points vertically projected on the horizontal plane, that is, the dihedral angle between the two vertical planes made through the two direction lines. It is one of the basic elements for determining the plane position of a point. The vertical angle (perpendicular angle) is the angle between the direction line from a certain point to a target point and the horizontal line in the same vertical plane and it is an element for determining the elevation position of a ground point.

2. 角度测量应用的仪器为光学经纬仪，应按正确的方法安置与使用。在仪器的使用上有诸多注意事项，应留心以提高测角的精度。经纬仪有四条主要轴线，它们之间应满足六项几何关系。当轴线之间应保证的几何关系遭到破坏时，应予以检验与校正，恢复轴线之间应有的几何关系，减小误差造成的影响。

2.The instrument used for angle measurement is optical theodolite, which shall be installed and used correctly. One should pay attention to many precautions in the use of the instrument to improve the accuracy of angle measurement. The theodolite has four main axes, which shall meet the six geometric relations. When the required geometric relationship among axes is damaged, it shall be checked and corrected to restore the proper geometric relationship among axes and reduce the impact of errors.

3. 水平角测量方法一般可根据观测目标的多少和工作要求的精度而定。常用的水平角测量方法有测回法和方向观测法。竖直角的测量应使用经纬仪的竖直度盘。

3.The horizontal angle measurement method can generally be determined according to the number of observation targets and the accuracy required by the work. The commonly used horizontal angle observation methods include observation set method and direction observation method. The vertical angle shall be measured with the vertical circle of the theodolite.

# 复习思考题

## Review Questions

### 一、填空题

### I. Fill in the Blanks

1.DJ 型光学经纬仪主要由______________、______________和______________三大部分组成。

1.DJ optical theodolite is mainly composed of three parts, ____________, ____________, and ____________.

2.DJ 型光学经纬仪的使用，包括______________、______________和______________三项基本操作。

2. The use of DJ optical theodolite includes three basic operations, ______________, ______________, and ______________.

3. 经纬仪的主要轴线有______________、______________、______________和______________。

3. The main axes of the theodolite include ______________, ______________, ______________, and ______________.

4. 常用的水平角测量方法有______________和______________两种。

4. The commonly used horizontal angle measurement methods include ____________ and ____________.

5. 水平角测量受多种误差的影响，主要有______________、______________和______________。

5. The measurement horizontal angle will be affected by various error, mainly including ______________, ______________ and ______________.

## 二、选择题

## II. Choice Questions

1. 经纬仪对中的目的是使仪器的中心（竖轴）与测站点位于同一（　）上。

1. The purpose of theodolite centering is to make the center (vertical axis) of the instrument on the same (　) as the observation station.

A. 水平线　　B. 铅垂线　　C. 倾斜线　　D. 水平面

A. horizontal line　　B. plumb line　　C. inclined line　　D. horizontal plane

2. 经纬仪安置时，整平的目的是使仪器的（　）。

2. When the theodolite is placed, the purpose of leveling is to (　).

A. 竖轴位于铅垂位置，水平度盘水平

A. make the vertical axis at the plumb line position, and the horizontal circle horizontal

B. 水准管气泡居中

B. make the level tube bubble centered

C. 竖盘指标处于正确位置

C. make the vertical circle index in the correct position

D. 圆水准器气泡居中

D. make the circular level bubble centered

3. 经纬仪视准轴检验和校正的目的是（　）。

3. The purpose of theodolite collimation axis inspection and calibration is to (　).

A. 使横轴垂直于竖轴

A. make the horizontal axis perpendicular to the vertical axis

B. 使视准轴垂直于横轴

B. make the collimation axis perpendicular to the horizontal axis

C. 使视准轴平行于水准管轴

C. make the collimation axis parallel to the level tube axis

D. 使视准轴平行于横轴

D. make the collimation axis parallel to the horizontal axis

4. 经纬仪的照准部水准管轴应（　）于仪器竖轴。

4. The alidade level tube axis of the theodolite shall (　) the vertical axis of the instrument.

A. 平行　　B. 垂直　　C. 倾斜　　D. 重合

A. be parallel to　　B. be perpendicular to　　C. deviate from　　D. coincide with

5. 竖直角的最大值为（　）。

5. The maximum value of a vertical angle is (　).

A. 90°　　B. 180°　　C. 270°　　D.360°

## 三、简答题

## III. Short Answer Questions

1. 什么是水平角？瞄准同一竖直面上高度不同的点，它们在水平度盘上的读数是否相同？为什么？

1. What is the horizontal angle? Are the readings on the horizontal circle the same when points at different heights on the same vertical plane are aimed at? And why?

2. 什么是竖直角？瞄准同一铅垂面内不同高度的点，它们在竖直度盘上的读数是否相同？为什么？

2. What is the vertical angle? Are the readings on the vertical circle the same when points at different heights on the same vertical plane are aimed at? And why?

3. 经纬仪有哪些主要轴线？各轴线之间应满足什么条件？

3. What are the main axes of the theodolite? What conditions should be met among the axes?

4. 什么是竖盘指标差？如何消除？

4. What is the index error of vertical circle? How to eliminate it?

4

# 小区域控制测量

# Small Area Control Survey

**学习目标：**

**Learning Objectives:**

通过本章的学习，了解控制测量的基本知识，导线测量的技术要求，三角测量的主要技术要求，三、四等水准测量的技术要求；熟悉导线测量的外业工作，三角测量的观测作业，三、四等水准测量的外业观测，三角高程测量的原理与技术要求；掌握导线测量内业计算方法，三角形网测量数据处理方法，三、四等水准测量的内业计算方法，常用三角高程测量方法。

Through the study of this chapter, to understand the basic knowledge of control survey and the technical requirements of traverse survey, triangular network survey, and third-order and fourth-order leveling; to get familiar with the field work of traverse survey, the observation of triangular network survey, the field observation of third-order and fourth-order leveling, and the principle and technical requirements of trigonometric leveling; to master the office computation method of traverse survey, the data processing method of triangular network survey, the office computation method of third-order and fourth-order leveling, and the commonly used method of trigonometric leveling.

**能力目标：**

**Ability Objectives:**

理解小区域控制测量（平面控制测量和高程控制测量）的基本原理和方法，能够进行导线测量，三、四等水准测量，三角高程测量的外业观测和内业计算。

Be able to understand the basic principles and methods of small area control survey (horizontal control survey and vertical control survey); to be able to carry out the field observation and office computation of traverse survey, third-order and fourth-order leveling, and trigonometric leveling.

# 4.1 控制测量概述

## 4.1 Overview of Control Survey

### 4.1.1 控制测量的概念

### 4.1.1 Concept of Control Survey

无论是测绘地形图还是施工放样，测量过程中都不可避免地会产生误差。为了限制测量误差的累积，保证测图和施工测量的精度和速度，测量工作必须采取正确的测量程序和方法，即遵循“先整体后局部，先控制后碎部”的原则。也就是说，测量工作必须先进行控制测量，同时，通过控制测量统一测量所用的坐标系统和高程系统。

Errors will inevitably occur during surveying and mapping of topographic maps and construction setting-out. In order to limit the accumulation of survey errors and ensure the accuracy and efficiency of mapping and construction survey, correct survey procedures and methods must be adopted, that is, following the principle of “overall survey before local survey, control survey before detail survey”. That is to say, the control survey must be carried out first to define the unified coordinate system and elevation system used for the survey work.

在测区内，按测量任务所要求的精度，测定一系列控制点的平面位置和高程，建立起测量控制网，作为各种测量的基础，这种测量工作称为控制测量；测定控制点平面位置的工作称为平面控制测量；测定控制点高程的工作称为高程控制测量。

In the survey area, it is called the control survey to measure and determine the plane positions and elevations of a series of control points according to the accuracy required by the survey task and then to establish a survey control network as the basis for various survey tasks; it is called the horizontal control survey to measure and determine the plane positions of control points; and it is called the vertical control survey to measure and determine the elevations of control points.

### 4.1.2 平面控制测量

### 4.1.2 Horizontal Control Survey

平面控制测量的主要方法有三角测量和导线测量。

The main methods of horizontal control survey are trigonometric survey and traverse survey.

#### （1）三角测量

#### (1) Trigonometric survey

①三角锁（网）。按要求在地面上选择一系列具有控制作用的控制点，组成互相连接的三角形，若三角形排列成条状，称为三角锁［图 4-1（a）］；若扩展成网状，称为三角网［图 4-1（b）］。

① Triangulation chain (network). A series of control points on the ground are selected as required to form interconnected triangles. If these triangles are arranged in strips, it is called a triangulation chain [Figure 4-1(a)]; if they are expanded into a mesh, it is called a triangulation network [Figure 4-1(b)].

②三角测量。三角测量是用精密仪器观测三角锁（网）中所有三角形的内角，并精确测定起始边的边长和方位角，然后根据三角公式解算出各点的坐标。

② Trigonometric survey. Trigonometric survey is to observe and measure the inner angles of all triangles in the triangulation chain (network), accurately measure the side length and azimuth of the starting edge, and then calculate the coordinates of each point according to the trigonometric formula.

在全国范围内统一建立的三角网，称为国家平面控制网。国家平面控制网按精度从高到低分为一等、二等、三等、四等四个等级。其中，一等三角网是国家平面控制网的骨干；二等三角网布设于一等三角锁环内，是国家平面控制网的全面基础；三等、四等三角网是二等三角网的进一步加密。

The united triangulation network established nationwide is called the national horizontal control network. The national horizontal control network is divided into four orders according to the accuracy from high to low. The first-order triangulation network is the backbone of the national horizontal control network; the second-order triangulation network, arranged inside the first-order triangulation network, is the comprehensive foundation of the national horizontal control network; the third-order and fourth-order triangulation networks are the further refinement of the second-order triangulation network.

### （2）导线测量

### (2) Traverse survey

将相邻控制点依次用直线相连而组成的折线称为导线，构成导线的控制点称为导线点。导线测量就是依次测量各导线边的水平距离以及相邻导线边的水平夹角，然后根据起算数据，推算各导线边的坐标方位角，从而求出各导线点的平面坐标。

The polyline formed by connecting adjacent control points in turn with straight lines are called a traverse. The control points that make up the traverses are called traverse points. The traverse survey is to measure the horizontal distance between traverse legs and the horizontal angles between adjacent traverse legs in turn, and then calculate the coordinate azimuth angles of various traverse legs according to the known data, so as to obtain the horizontal coordinates of various traverse points.

导线测量是建立小地区平面控制网常用的一种方法，特别是在地物分布较复杂的建筑区、视线障碍较多的隐蔽区和带状地区，多采用导线测量的方法。

Traverse survey is a common method to

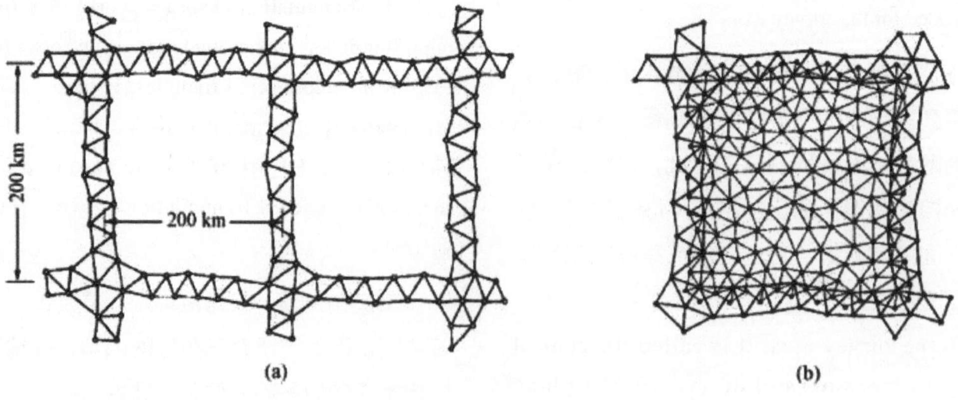


图4-1　三角锁（网）

Figure 4-1 Triangulation Network

establish the horizontal control network in small areas, especially in building areas distributed with complex surface features, hidden areas and strip areas with many sight obstacles.

## 4.1.3 高程控制测量

### 4.1.3 Vertical Control Survey

高程控制测量的方法主要有水准测量和三角高程测量。高程控制测量精度等级的划分，依次为二等、三等、四等、五等。各等级高程控制宜采用水准测量，四等及以下等级可采用电磁波测距三角高程测量；五等也可采用 GPS 拟合高程测量。首级高程控制网的等级，应根据工程规模、控制网的用途和精度要求合理选择。首级网应布设成环形网，加密网宜布设成附合路线或结点网。测区的高程系统，宜采用 1985 年国家高程基准。在已有高程控制网的地区测量时，可沿用原有的高程系统；当小测区联测有困难时，也可采用假定高程系统。高程控制点间的距离，一般地区应为 1 ~ 3 km，工业厂区、城镇建筑区宜小于 1 km。但一个测区及周围至少应有三个高程控制点。

Vertical control survey methods mainly include leveling and trigonometric leveling. There are generally four orders of vertical control survey accuracy, namely, second-order, third-order, fourth-order and fifth-order. Leveling should be adopted for each order of vertical control, generally trigonometric leveling with electromagnetic distance measurement for fourth-order or below, and GPS leveling fitting for fifth-order. The order of the primary vertical control network shall be reasonably determined according to the scale of work, the purpose and precision requirements of the control network. The primary vertical control network shall be arranged as a ring network, while the densified control network should be arranged as a connecting traverse or junction point traverse network. The 1985 national elevation datum should be adopted for the elevation system of the survey area. The original elevation system can be used for surveying in areas with existing vertical control network; when it is difficult to conduct connection survey in small survey areas, the assumed elevation system can also be used. The distance between vertical control points shall be 1 to 3 km in general areas, and less than 1 km in industrial plant areas and urban built-up areas. However, there shall be at least three vertical control points in and around a survey area.

## 4.1.4 小区域平面控制测量

### 4.1.4 Small Area Horizontal Control Survey

在小区域（面积 ≤ 15 $km^2$）内建立的平面控制网，称为小区域平面控制网。小区域平面控制网应尽可能与当地已经建立的国家或城市控制网联测，并以国家或城市控制网的数据作为起算和校核标准。如果测区范围附近没有合适的高等级控制点，或附近有合适的高等级控制点但不方便联测，也可以建立测区独立控制网。

The horizontal control network established in a small area (area ≤ 15 $km^2$) is called the small area horizontal control network. The small area horizontal control network shall be connected with the established national or urban control network as far as possible for connection survey, and the data of the national or urban control network shall be used as the calculation and check standard. If there are no suitable high-order control points near the survey area, or there are suitable high-order control points nearby but it is not convenient for connection survey, an independent control network can also be established in the survey area.

小区域平面控制网也应由高级到低级分级建立。测区范围内建立最高一级的控制网，称为首级控制网；最低一级的直接为测图而建立的控制网，称为图根控制网。首级控制与图根控制的关系见表 4-1。

Small area horizontal control network shall also be established with orders from high to low. The control network with the highest order established in a survey area is called the primary control network; the control network with the lowest order established for mapping is called the mapping control network. The relationship between the primary control network and the mapping control network is shown in Table 4-1.

直接用于地形测图的控制点称为图根控制点，简称图根点。图根点位置的测定工作，称为图根控制测量。图根点的密度取决于测图比例尺和地形的复杂程度，具体应符合表 4-2 的规定。

The control points directly used for topographic mapping are called mapping control points. The determination of mapping control points is called mapping control survey. The density of mapping control points depends on the mapping scale and the complexity of the terrain, which shall comply with the provisions of Table 4-2.

表 4-1　首级控制与图根控制的关系

Table 4-1 Relationship between Primary Control and Mapping Control

| 测区面积/$km^2$<br>Survey area/$km^2$ | 首级控制<br>Primary control | 图根控制<br>Mapping control |
|---|---|---|
| 1～10 | 一级小三角或一级导线<br>First-order minor triangulation or first-order traverse | 两级图根<br>Second-order mapping |
| 0.5～2 | 二级小三角或二级导线<br>Second-order minor triangulation or second-order traverse | 两级图根<br>Second-order mapping |
| 0.5以下<br>Below 0.5 | 图根控制<br>Mapping control | |

表 4-2　图根点密度

Table 4-2 Density of Mapping Control Points

| 测图比例尺<br>Mapping scale | 每平方千米图根点数量/个<br>Number of mapping point in 1 $km^2$ |
|---|---|
| 1：5 000 | 5 |
| 1：2 000 | 15 |
| 1：1 000 | 50 |
| 1：500 | 150 |

# 4.2 导线测量

## 4.2 Traverse Survey

### 4.2.1 导线的布设形式

### 4.2.1 Layout Form of Traverse

根据测区的不同情况和具体要求，导线可按下列三种形式进行布设。

According to different conditions and specific requirements of the survey area, the traverse can be laid in the following three forms.

#### （1）闭合导线

#### (1) Closed traverse

闭合导线是从一个已知点 $B$ 出发，经过若干个导线点1、2、3、4、……又回到原已知点 $B$ 上，形成一个闭合多边形，如图 4-2 所示。

A closed traverse starts from a known point $B$, passes through several traverse points 1, 2, 3, 4, ..., and returns to the original known point $B$ to form a closed polygon, as shown in Figure 4-2.

#### （2）附合导线

#### (2) Connecting traverse

附合导线是从一条已知边的一个已知点出发，经过一系列导线点，最后附合到另一条已知边的一个已知点上，如图 4-3 所示。

A connecting traverse starts from a known point on a known edge, passes through a series of traverse points, and finally attaches to a known point on another known edge, as shown in Figure 4-3.

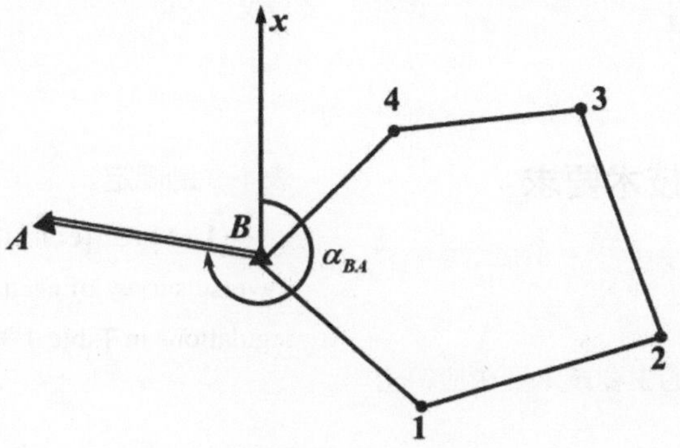


图4-2 闭合导线

Figure 4-2 Closed Traverse

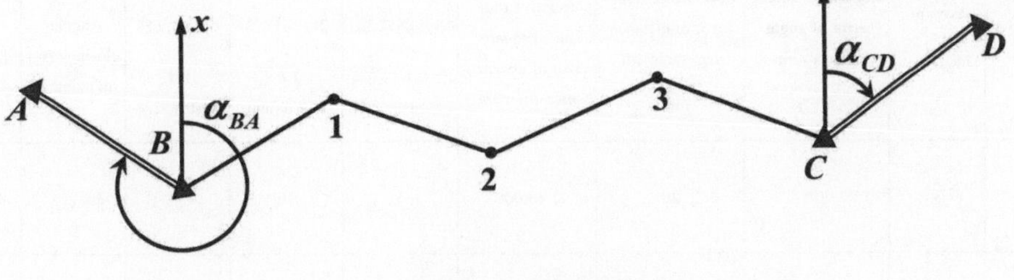


图4-3 附合导线

Figure 4-3 Connecting Traverse

### (3) 支导线

(3) Open traverse

支导线是从一个已知控制点和一个已知方向出发，既不附合到另一已知控制点上，也不回到原起点上，如图 4-4 所示。

An open traverse starts from a known control point and a known direction, and is neither attached to another known control point nor returns to the original starting point, as shown in Figure 4-4.

支导线只具有必要的起始数据，缺少对观测成果的检核，因此，仅用于图根控制测量，而且一条导线上布设的导线点一般不得超过四个。

The open traverse only has necessary initial data and lacks the verification of observation results. Therefore, it is only used for mapping control survey, and the number of traverse points laid on one traverse shall not exceed four.

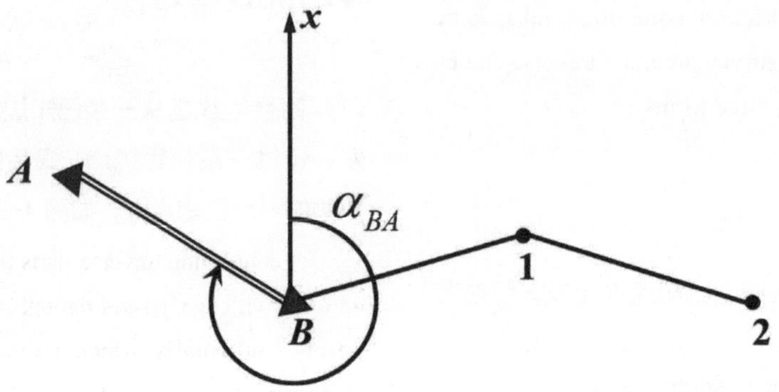


图4-4 支导线

Figure 4-4 Open Traverse

## 4.2.2 导线测量的技术要求

## 4.2.2 Technical Requirements for Traverse Survey

①各等级导线测量的主要技术要求应符合表 4-3 的规定。

① Main technical requirements for the traverse survey of each order shall comply with the regulations in Table 4-3.

表 4-3 导线测量的主要技术要求

Table 4-3 Main Technical Requirements for Traverse Survey

| 等级 Class | 导线长度 Traverse length /km | 平均边长 Average side length /km | 测角中误差 Root mean square error of angle observation /(″) | 测距中误差 Root mean square error of distance measurement / mm | 测距相对中误差 Relative root mean square error of distance measurement | 测回数 Number of observation sets | | | 方位角闭合差 Angular closing error of azimuth /(″) | 导线全长相对闭合差 Relative total length closing error of traverse |
|---|---|---|---|---|---|---|---|---|---|---|
| | | | | | | 1″级仪器 1″ instrument | 2″级仪器 2″ instrument | 6″级仪器 6″ instrument | | |
| 三等 Third-order | 14 | 3 | 1. 8 | 20 | 1/15 000 | 6 | 10 | — | $3.6\sqrt{n}$ | ≤1/55 000 |
| 四等 Fourth-order | 9 | 1.5 | 2. 5 | 18 | 1/80 000 | 4 | 6 | — | $5\sqrt{n}$ | ≤1/35 000 |

续表

| 等级<br>Class | 导线长度<br>Traverse length<br>/km | 平均边长<br>Average side length<br>/km | 测角中误差<br>Root mean square error of angle observation<br>/(″) | 测距中误差<br>Root mean square error of distance measurement<br>/ mm | 测距相对中误差<br>Relative root mean square error of distance measurement | 测回数<br>Number of observation sets | | | 方位角闭合差<br>Angular closing error of azimuth<br>/(″) | 导线全长相对闭合差<br>Relative total length closing error of traverse |
|---|---|---|---|---|---|---|---|---|---|---|
| | | | | | | 1″级仪器<br>1″ instrument | 2″级仪器<br>2″ instrument | 6″级仪器<br>6″ instrument | | |
| 一级<br>First-class | 4 | 0. 5 | 5 | 15 | 1/30 000 | — | 2 | 4 | $10\sqrt{n}$ | ≤1/15 000 |
| 二级<br>Second-class | 2. 4 | 0. 25 | 8 | 15 | 1/14 000 | — | 1 | 3 | $16\sqrt{n}$ | ≤1/10 000 |
| 三级<br>Third-class | 1. 2 | 0. 1 | 12 | 15 | 1/7 000 | — | 1 | 2 | $24\sqrt{n}$ | ≤1/5 000 |

注：1.表中*n*为测站数。

Note: 1. In the above table, "*n*" stands for the number of observation stations.

2.当测区测图的最大比例尺为1∶1 000时，一、二、三级导线的导线长度、平均边长可适当放长，但最大长度不应大于表中规定相应长度的两倍。

2. When the maximum scale of mapping in the survey area is 1∶1000, the length and average leg length of the first-class, second-class and third-class traverses can be appropriately lengthened. However, the maximum length shall not be greater than 2 times the corresponding length specified in the table.

②当导线平均边长较短时，应控制导线边数不超过表 4-3 相应等级导线长度和平均边长算得的边数；当导线长度小于表 4-3 规定长度的 1/3 时，导线全长的绝对闭合差不应大于 13 cm。

② When the average length of the traverse legs is short, the number of traverse legs shall not exceed the number of legs calculated by the length and average leg length of the traverse of the corresponding class stated in Table 4-3. When the length of the traverse is less than 1/3 of the length specified in Table 4-3, the absolute total length closing error of the traverse shall not be greater than 13 cm.

③导线网中，结点与结点、结点与高级点之间的导线段长度不应大于表 4-3 中相应等级规定长度的 70%。

③ In the traverse network, the length of the traverse sections between junction points and between junction points and higher-order points shall not be greater than 70% of the length specified in Table 4-3.

④导线网的布设应符合下列规定。

④ The layout of traverse network shall comply with the provisions below.

a. 导线网用作测区的首级控制时，应布设成环形网，且宜联测两个已知方向。

a. When the traverse network is used as the primary control of the survey area, it shall be arranged as a ring network, and two known directions should be selected for connection survey.

b. 加密网可采用单一附合导线或结点导线网形式。

b. The densified network can be in the form of a single connecting traverse or a junction point traverse network.

c. 结点间或结点与已知点间的导线段宜布

设成直伸形状，相邻边长不宜相差过大，网内不同环节上的点也不宜相距过近。

c. The traverse sections between the junction points or between the junction points and the known points shall be arranged in a straight-out shape, the length difference of adjacent traverse legs shall not be too large, and the junction points in the network shall not be too close to each other.

# 4.3 导线测量的外业工作

## 4.3 Field Work of Traverse Survey

导线测量的外业工作包括踏勘选点、测角、量边、定向。其工作内容如下。

The field work of traverse survey includes point selection, angle measurement, leg length measurement and orientation. The specific work contents are described as follows:

### 4.3.1 踏勘选点

### 4.3.1 Point Selection

踏勘选点之前，应先到有关部门收集原有地形图、高一级控制点的坐标和高程，以及这些已知点的位置详图。然后按坐标把已知点展绘在原有的地形图上，在图上规划导线的布设方案。最后，带上所规划的导线网图，到实地选定各点点位并建立标志。

Before point selection, collect the original topographic map, the coordinates and elevations of the higher-order control points, and the location details of these known points from relevant departments. Then, draw the known points on the original topographic map according to the coordinates, and plan the layout scheme of the traverse on the map. Finally, take the planned traverse network map to the site, select the points and establish signs.

导线点位的选定应符合下列规定。

The selection of traverse points shall comply with the provisions below.

①点位应选在土质坚实、稳固可靠、便于保存的地方，视野应相对开阔，便于加密、扩展和寻找。

① The points shall be located in stable and reliable places with solid soil, convenient storage, and broad vision conductive to densification, expansion and searching.

②相邻点之间应通视良好，其视线距障碍物的距离，三等、四等不宜小于 1.5 m，四等以下宜保证便于观测，以不受旁折光的影响为原则。

② Adjacent points shall have good intervisibility, the distance between their sight lines and obstacles shall not be less than 1.5 m for third-order and fourth-order survey, and easy observation without influence due to lateral refraction should be ensured for the lower-order survey.

③当采用电磁波测距时，相邻点之间视线应避开烟囱、散热塔及强电磁场。

③ When electromagnetic distance measurement is adopted, the line of sight between adjacent points shall be far from chimneys, cooling towers and strong electromagnetic fields.

④相邻两点之间的视线倾角不宜过大。

④ The inclination of the line of sight between two adjacent points should not be too large.

⑤充分利用已有控制点。

⑤ Make full use of the existing control points.

⑥导线点应有足够的密度，分布要均匀，以便于控制整个测区。

⑥ The traverse points shall be properly densified and evenly distributed, so as to control the whole survey area.

⑦导线边长应大致相等，尽量避免相邻边长相差悬殊，以保证和提高测角精度。

⑦ The traverse legs shall have roughly the same length and large length difference between two adjacent traverse legs shall be avoided as far as possible, so as to ensure and improve the angle measurement accuracy.

导线点选定后，应用明显的标志固定下来，通常是用一木桩打入土中，桩顶高出地面 1~2 cm, 并在桩顶钉一小钉，作为临时标志，如图 4-5(a) 所示。当导线点选择在水泥、沥青等坚硬地面时，可直接钉一钢钉作为标志，需要长期保存使用的导线点，应埋设混凝土桩，桩顶刻“十”字，作为永久性标志，如图 4-5(b) 所示。导线点选定后，应进行统一编号，并绘制导线线路草图，标记点位。

Traverse points, after selected and determined, shall be fixed with obvious marks, usually by driving wooden piles into the earth, with the pile top 1-2 cm above the ground, and a small pin nailed on the pile top as a temporary mark, as shown in Figure 4-5(a). When the traverse point is selected on hard ground such as cement or asphalt floor, a steel nail can be directly nailed as a mark. At the traverse point that needs to be preserved and used for a long time, a concrete pile shall be embedded as a permanent mark, with a “cross” engraved on the pile top, as shown in Figure 4-5(b). The selected traverse points shall be numbered uniformly, and the traverse sketch and the location of the points shall be marked.

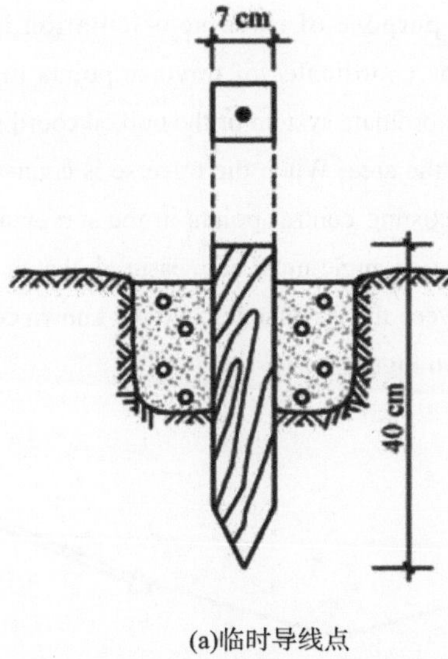


(a)临时导线点

(a) Temporary Traverse Points

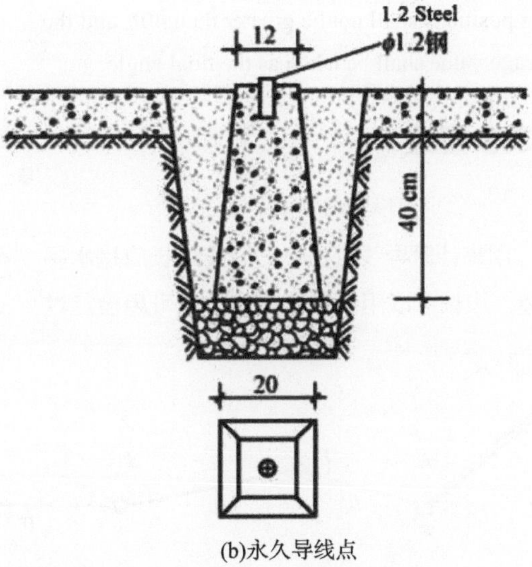


(b)永久导线点

(b) Permanent Traverse Points

图4-5　导线点埋设

Figure 4-5 Embedding of Traverse Points

### 4.3.2 测角

### 4.3.2 Angle Measurement

导线转折角有左、右之分，以导线为界，沿前进方向向左侧的角称为左角；沿前进方向向右侧的角称为右角。在附合导线中一般测量其左角，在闭合导线中一般测量其内角。闭合导线若按逆时针方向编号，其内角即为左角；反之，则为右角。对于图根导线，一般用经纬仪观测一个测回，盘左、盘右测得角度之差不得大于 40″，并取平均值作为最后角度。

The traverse turning angle is divided into left and right angles. The angle to the left of the traverse along the forward direction is called the left angle, and the angle to the right of the traverse along the forward direction is called the right angle. Generally, the left angle is measured for the connecting traverse and the inner angle is measured for the closed traverse. If the angles are numbered counterclockwise for the closed traverse, the inner angles are left angles and the rest are right angles. For the mapping traverse, a theodolite is generally used to perform one observation set. The difference between the angles measured in the face left and right positions shall not be greater than 40″, and the average value shall be taken as the final angle.

### 4.3.3 量边

### 4.3.3 Leg Length Measurement

用来计算导线点坐标的导线边长应是水平距离。边长可以用全站仪观测，也可用检定过的钢尺丈量。对于等级导线，应按规范进行精密测距；对于图根导线，若用钢尺量距，可以往、返各丈量 1 次，也可以同一方向丈量 2 次，取其平均值，其相对误差不大于 1/3 000。

The traverse leg length used to calculate the coordinates of the traverse points shall be the horizontal distance. The leg length can be measured with either a total station or a calibrated steel tape. For graded traverses, precise distance measurement shall be carried out according to the specifications. For mapping traverses, if a steel tape is used to measure the distance, either a forward and backward reservation or double reservations in one direction can be carried out, and the average value shall be taken, with a relative error of not more than 1/3 000.

### 4.3.4 定向

### 4.3.4 Orientation

导线定向的目的，是使导线点的坐标纳入国家坐标系或该地区的统一坐标系中。当导线与测区已有控制点连接时，必须测出连接角，即导线边与已知边发生联系的角，如图 4-6 所示。

The purpose of traverse orientation is to include the coordinates of traverse points in the national coordinate system or the unified coordinate system of the area. When the traverse is connected with the existing control points in the survey area, the connection angle must be measured, that is, the angle between the traverse leg and the known edge, as shown in Figure 4-6.

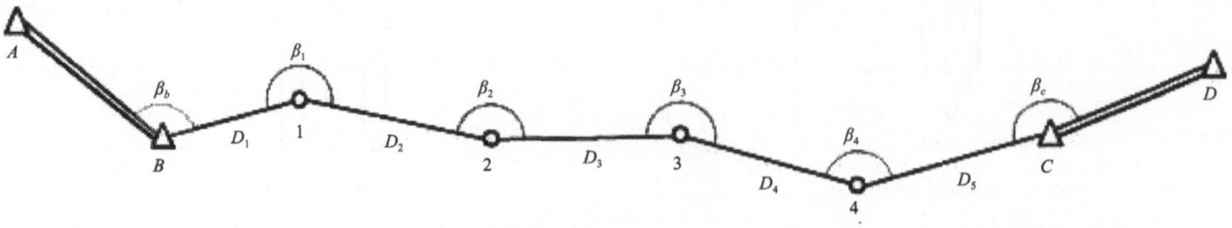


图4-6　连接测量示意图

Figure 4-6 Schematic Diagram of Connection Survey

# 4.4 导线测量的内业计算

## 4.4 Office Computation of Traverse Survey

导线测量的内业计算是在导线测量外业工作完成后，合理地进行各种误差的计算和调整，计算出各导线点坐标的工作。在进行导线测量内业计算之前，一是要全面检查外业观测数据有无遗漏，记录、计算是否正确，结果是否符合精度要求。当发现记录、计算有错时，不要改动原始数据，要认真反复校核。二是要根据已知数据和观测结果绘制外业成果注记图。当确定外业结果符合要求后，才可进行内业的计算。

The office computation of traverse survey is to reasonably calculate and adjust various errors and calculate the coordinates of each traverse point after the field work of traverse survey is completed. Before the office computation of traverse survey, first comprehensively check whether there is any omission in the field observation data, whether the records and calculations are correct, and whether the results meet the accuracy requirements. When errors are found in records and calculations, the original data shall not be changed, and shall be checked carefully and repeatedly. Second, prepare the annotated map according to the known data and observation results. The office computation can be carried out only after it is determined that the field work results meet the requirements.

### 4.4.1 测量坐标的正算与反算

### 4.4.1 Forward and Backward Calculation of Survey Coordinates

#### (1) 坐标正算

#### (1) Forward calculation of coordinates

①定义：设已知点 $A$ 的坐标为（$x_A$，$y_A$），测得 $A$、$B$ 之间的距离 $D_{AB}$ 及方位角 $\alpha_{AB}$，求待定点 $B$ 的坐标（$x_B$，$y_B$）。

① Definition: Assume that the coordinates of the known point $A$ are ($x_A$, $y_A$), measure the distance $D_{AB}$ between $A$ and $B$ and the azimuth angle $\alpha_{AB}$ and calculate the coordinates of point $B$ to be determined ($x_B$, $y_B$).

②坐标正算如图 4-7 所示。计算公式为

② The forward calculation of coordinates is shown in Figure 4-7. The calculation formula is as follows

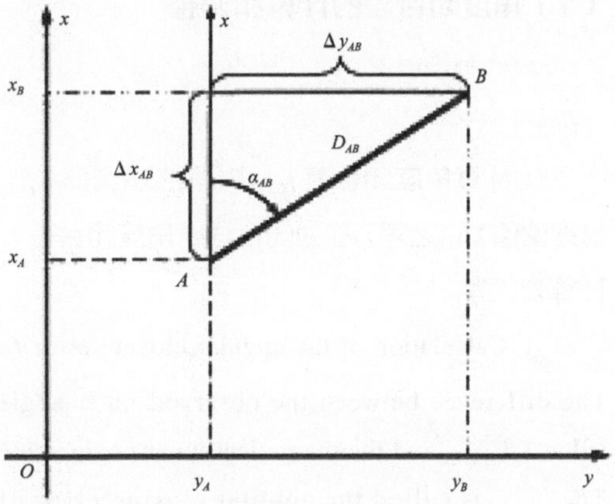


图4-7　坐标正算

Figure 4-7 Coordinate Forward Calculation

$$\Delta x=D\cos\alpha_{AB};\quad \Delta y=D\sin\alpha_{AB}$$

$$x_B=x_A+\Delta x=x_A+D\cos\alpha_{AB};$$

$$y_B=y_A+\Delta y=y_A+D\sin\alpha_{AB}$$

#### (2) 坐标反算

#### (2) Backward calculation of coordinates

定义：已知 $A$、$B$ 两点的坐标为（$x_A$，$y_A$）和（$x_B$，$y_B$），计算两点之间的距离及方位角。

Definition: Given that the coordinates of

points $A$ and $B$ are $(x_A, y_A)$ and $(x_B, y_B)$, calculate the distance between the two points and the azimuth angle.

$$\tan\alpha_{AB} \frac{\Delta y}{\Delta x} = \frac{y_B - y_A}{x_B - x_A} \quad (4.1)$$

$$D = \sqrt{\Delta x^2 + \Delta y^2} = \sqrt{(x_B - x_A)^2 + (y_B - y_A)^2} \quad (4.2)$$

## 4.4.2 闭合导线的内业计算

## 4.4.2 Office Computation of Closed traverse

### （1）角度闭合差的计算和调整

### (1) Calculation and adjustment of angular closing error

①计算角度闭合差$f_\beta$：内角的观测值$\Sigma\beta_{测}$与理论值$\Sigma\beta_{理}$之差$f_\beta$称为闭合导线角度闭合差。计算公式为

① Calculation of the angular closing error $f_\beta$: The difference between the observed inner angle value $\Sigma\beta_{Observed}$ and the theoretical inner angle value $\Sigma\beta_{Theoretical}$ is called the angular closing error of closed traverse $f_\beta$. The calculation formula is as follows.

$$f_\beta = \Sigma\beta_{测} - \Sigma\beta_{理} = \Sigma\beta_{测} - (n-2)\cdot 180° \quad (4.3)$$

$$f_\beta = \Sigma\beta_{Observed} - \Sigma\beta_{Theoretical} = \Sigma\beta_{Observed} - (n-2)\cdot 180° \quad (4.3)$$

②计算角度闭合差允许值$f_{允}$：按导线转折角观测和限差表的规定计算。

② Calculation of allowable angular closing error $f_{Allowed}$ : Follow the traverse turning angle observation and tolerance provisions.

③判断精度：当$f_\beta \leqslant f_{允}$时，满足精度要求，超过则重测。

③ Judgment of precision: When $f_\beta \leqslant f_{Allowed}$, the precision requirement is met; otherwise, resurvey is required.

④计算角度改正数方法：闭合差按相反符号平均分配到各角上；当$f_\beta$不能整除时，余数分在短边两侧的角上。

④ Calculation of the angle correction: The closing error is evenly distributed to each angle with the opposite sign; when $f_\beta$ cannot be divided evenly, the remainder is divided to angles on both sides of the short leg.

### （2）导线边方位角的计算

### (2) Calculation of traverse leg azimuth

可根据第一条边的方位角和调整后的内角（左角），推算其他各边的方位角，其计算公式为

According to the azimuth angle of the first leg and the adjusted inner angle (left angle), the azimuth angles of other legs can be calculated by the following formula.

$$\alpha_{前} = \alpha_{后} + \beta_{左} - 180° \quad （左角） \quad (4.4)$$

$$\alpha_{Front} = \alpha_{Rear} + \beta_{Left} - 180° \quad (\text{left angle}) \quad (4.4)$$

$$\alpha_{前} = \alpha_{后} - \beta_{右} + 180° \quad （右角） \quad (4.5)$$

$$\alpha_{Front} = \alpha_{Rear} - \beta_{Right} + 180° \quad (\text{right angle}) \quad (4.5)$$

式中，$\beta_{左}$是改正后的左角；

Where, $\beta_{Left}$ is the corrected left angle;

$\beta_{右}$是改正后的右角。

$\beta_{Right}$ is the corrected right angle.

当采用式 (4.5) 算得的值超过 360° 时，应减去 360° 。由最后一边的方位角推算而得第一边的方位角，其值应等于它的起始值，如不等于，则表明计算有错误。

When the value calculated by Formula (4.5)

exceeds 360°, it shall be subtracted by 360°. The azimuth of the first leg is deduced from the azimuth of the last leg and it shall be equal to its starting value. If it is not equal, it indicates that there is an error in the calculation.

### (3) 坐标增量计算及坐标增量闭合差的调整

### (3) Calculation of coordinate increment and adjustment of coordinate increment closing error

①坐标增量的计算。按坐标正算公式计算各边的坐标增量，其计算公式如下。

① Calculation of coordinate increment. Calculate the coordinate increment of each leg according to the forward calculation formula of coordinates as follows.

$$\Delta x = D\cos a_{AB};\ \Delta y = D\sin a_{AB} \tag{4.6}$$

计算位数取到 cm。

Calculate the number of digits to cm.

②坐标增量闭合差的计算。

② Calculation of increment closing error of vertical coordinate.

$$f_x = \Sigma\Delta x_{测} \tag{4.7}$$

$$f_x = \Sigma\Delta x_{\text{Observed}} \tag{4.7}$$

$$f_y = \Sigma\Delta y_{测} \tag{4.8}$$

$$f_y = \Sigma\Delta y_{\text{Observed}} \tag{4.8}$$

式中，$f_x$ 是纵坐标增量闭合差；

Where, $f_x$ is the increment closing error of vertical coordinate;

$f_y$ 是横坐标增量闭合差。

$f_y$ is the increment closing error of horizontal coordinate.

③导线全长闭合差的计算。

③ Calculation of total length closing error of traverse.

$$f = \sqrt{f_x^2 + f_y^2} \tag{4.9}$$

④导线相对闭合差。

④ Relative closing error of traverse.

$$K = \frac{f_s}{\Sigma D} = \frac{1}{\Sigma D / f_s} = \frac{1}{N} \tag{4.10}$$

在通常情况下，图根导线的 $K$ 值不应超过 1/2 000，困难地区也不应超过 1/1 000。

The $K$ value of the mapping control traverse shall not exceed 1/2 000 in general cases, and shall not exceed 1/1 000 in areas with survey difficulties.

⑤计算坐标增量改正值。

⑤ Calculation of the coordinate increment correction.

$$v_{xi} = -\frac{f_x}{\Sigma D} D_i;\ v_{yi} = -\frac{f_y}{\Sigma D} D_i \tag{4.11}$$

式中，$v_{xi}$、$v_{yi}$ 是第 $i$ 条边纵、横坐标增量改正值；$D_i$ 是第 $i$ 条边的边长。

Where, $v_{xi}$, $v_{yi}$ is the increment correction value of the vertical and horizontal coordinates of the ith leg; $D_i$ is the length of the ith leg.

### (4) 计算每一导线点的坐标

### (4) Calculation of the coordinates of each traverse point

根据起点的已知坐标及调整之后的坐标增量，逐一推求。算完最后一点，还要再推算起点的坐标以作检核。

Calculate one by one according to the known coordinates of the starting point and the adjusted coordinate increment. After calculating the last point, deduce the coordinates of the starting point for verification.

闭合导线的内业计算算例见图 4-8。闭合导线计算见表 4-4。

An example of office computation of the

closed traverse is shown in Figure 4-8. The calculation of the closed traverse is shown in Table 4-4 .

闭合导线坐标推算时，从已知点 $A$ 开始，推算出 $B$、$C$、$D$ 点的坐标，最后推算回到 $A$ 点，应与原来 $A$ 点坐标相同，作为推算坐标的检核。

When calculating the coordinates of the closed traverse, deduce the coordinates of points $B$, $C$ and $D$ based on the known data of the point B, and finally deduce backward the coordinates of point $A$, which shall be the same as the original value of point $A$ , as a check of the deduced coordinates.

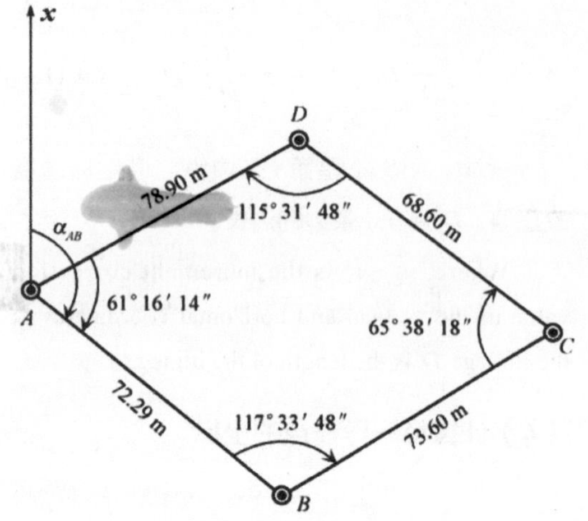


图4-8 闭合导线算例

Figure 4-8 Example of Closed Traverse Calculation

## 4.4.3 附合导线的内业计算

## 4.4.3 Office Computation of Connecting Traverse

图 4-9 所示为一附合导线，它的起点 1 和终点 $n$ 分别与高一级的控制点 $A$、$B$ 和 $C$、$D$ 连接，后者的坐标为已知，可按坐标反算公式计算起始边和终了边的方位角，即

Figure 4-9 shows a connecting traverse, with its starting point 1 and ending point $n$ connected to higher-order control points $A$ and $B$, $C$ and $D$ respectively. The coordinates of the latter points are known, and the azimuth angles of the starting and ending legs can be calculated according to the inverse calculation formula of the coordinate, i.e.

$$\alpha_{AB}=\arctan\frac{y_B-y_A}{x_B-x_A};\qquad \alpha_{CD}=\arctan\frac{y_D-y_C}{x_D-x_C}\tag{4.12}$$

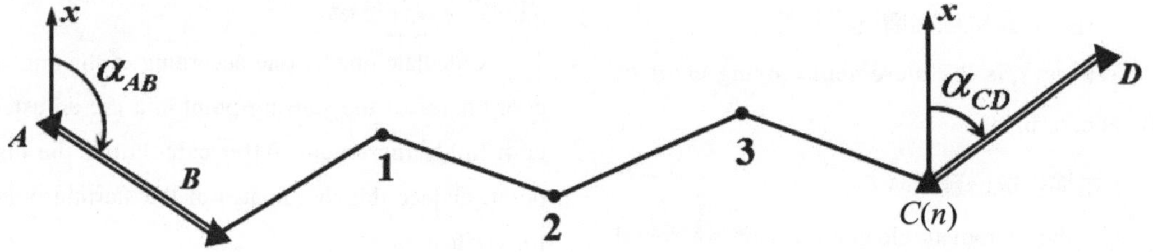


图4-9 附合导线计算

Figure 4-9 Calculation of Connecting Traverse

## 表 4-4　闭合导线计算

Table 4-4 Calculation of Closed Traverse

计算者：　　　　　　　　检查者：

Calculated by:　　　　　　Checked by:

| 点号 Point number | 观测角 Observation angle | 改正数 Correction | 改正后角值 Corrected rear angle | 坐标方位角 $\alpha$ Coordinate azimuth $\alpha$ | 距离/m Distance /m | 纵坐标增量 $\Delta x$ Vertical coordinate increment $\Delta x$ | | | 横坐标增量 $\Delta y$ Horizontal coordinate increment $\Delta y$ | | | 坐标值 Coordinate value | |
|---|---|---|---|---|---|---|---|---|---|---|---|---|---|
| | | | | | | 计算值/m Calculated value /m | 改正数/cm Correction /cm | 改正后/m Corrected value /m | 计算值/m Calculated value /m | 改正数/cm Correction /cm | 改正后/m Corrected value /m | *x*/m | *y*/m |
| 1 | 2 | 3 | 4 | 5 | 6 | 7 | 8 | 9 | 10 | 11 | 12 | 13 | 14 |
| A | | | | | | | | | | | | 500. 00 | 500. 00 |
| | | | | 135° 24′ 00″ | 72. 29 | −51.47 | +1 | −51. 46 | 50. 76 | +3 | 50. 79 | | |
| B | 117° 33′ 48″ | −2″ | 117° 33′ 46″ | | | | | | | | | 448. 54 | 550.79 |
| | | | | 72° 57′ 46″ | 73. 60 | 21. 56 | +1 | 21.57 | 70. 37 | +3 | 70. 40 | | |
| C | 65° 38′ 18″ | −2″ | 65° 38′ 16″ | | | | | | | | | 470. 11 | 621. 19 |
| | | | | 318° 36′ 02″ | 68. 60 | 51.46 | 0 | 51.46 | −45.37 | +3 | −45.34 | | |
| D | 115° 31′ 48″ | −2″ | 115° 31′ 46″ | | | | | | | | | 521. 57 | 575.85 |
| | | | | 254° 07′ 48″ | 78. 90 | −21. 58 | +1 | -21. 57 | −75. 89 | +4 | −75.85 | | |
| A | 61° 16′ 14″ | −2″ | 61° 16′ 12″ | | | | | | | | | 500. 00 | 500. 00 |
| | | | | 135° 24′ 00″ | | | | | | | | | |
| Σ | 360° 00′ 08″ | −8″ | 360° 00′ 00″ | | 293.39 | −0. 03 | +3 | 0. 00 | −0. 13 | +13 | 0. 00 | | |
| 辅助计算 Auxiliary calculation | $f_\beta=\Sigma\beta_{测}-\Sigma\beta_{理}=8''$　$f_{\beta允}=\pm 60\sqrt{0.293}=32''$<br>$f_\beta=\Sigma\beta_{Observed}-\Sigma\beta_{Theoretical}=8''$　$f_{\beta Allowed}=\pm 60\sqrt{0.293}=32''$<br>$f_x=\Sigma x_{测}=-0.03$ m；$f_x=\Sigma\Delta y_{测}=-0.13$ m；$f_D=\sqrt{f_x^2+f_y^2}=0.13$ m；$K=\frac{f_D}{\Sigma_D}\approx\frac{1}{2\,257}$<br>$fx=\Sigma x_{Observed}=-0.03$ m；$fx=\Sigma\Delta y_{Observed}=-0.13$ m；$f_D=\sqrt{f_x^2+f_y^2}=0.13$ m；$K=\frac{f_D}{\Sigma_D}\approx\frac{1}{2\,257}$ | | | | | | | | | | | | |

附合导线与闭合导线的计算步骤基本相同，但几何条件不同，角度闭合差和坐标增量闭合差的计算就有所不同。

The calculation steps of the connecting traverse and closed traverse are basically the same, but the calculation of the angular closing error and coordinate increment closing error is different due to different geometric conditions.

### （1）角度闭合差的计算

(1) Calculation of angular closing error

根据两端方向已知的特点，可由导线起始边的方位角 $\alpha_{AB}$ 和 $\beta_i$，推算得出终了边的方位角 $\alpha'_{CD}$，即

According to the known characteristics of the directions at both ends, the azimuth angle $\alpha'_{CD}$ of the final leg can be deduced from the azimuth angle $\alpha_{AB}$ and left angle $\beta_i$ of the starting leg of the traverse, i.e.

$$\alpha'_{CD}=\alpha_{AB}+\Sigma\beta_i\pm n\cdot 180° \qquad (4.13)$$

附合导线的角度闭合差为

Angular closing error of connecting traverse is

$$f_\beta=\alpha'_{CD}-\alpha_{CD} \qquad (4.14)$$

$$f_\beta=\alpha_{AB}+\Sigma\beta_i\pm n\cdot 180°-\alpha_{CD} \qquad (4.15)$$

### （2）坐标增量闭合差的计算

(2) Calculation of increment closing error of vertical coordinate

$$f_x=\Sigma\Delta x_{测}-(x_C-x_B) \qquad (4.16)$$

$$f_x=\Sigma\Delta x_{Observed}-(x_C-x_B) \qquad (4.16)$$

$$f_y=\Sigma\Delta y_{测}-(y_C-y_B) \qquad (4.17)$$

$$f_y=\Sigma\Delta y_{Observed}-(y_C-y_B) \qquad (4.17)$$

附合导线各导线点坐标计算的其他内容同闭合导线。

Other contents of the coordinate calculation of each traverse point of the connecting traverse are the same as those of the closed traverse.

附合导线内业计算算例如图 4-10 所示。

An example of office calculation of the connecting traverse is shown in Figure 4-10.

附合导线坐标计算见表 4-5。

The coordinate calculation of the connecting traverse is shown in Table 4-5 .

## 4.4.4 支导线计算

## 4.4.4 Calculation of Open Traverse

支导线中没有检核条件，因此没有闭合差产生，导线转折角和计算的坐标增量均不需要进行改正。支导线的计算步骤如下。

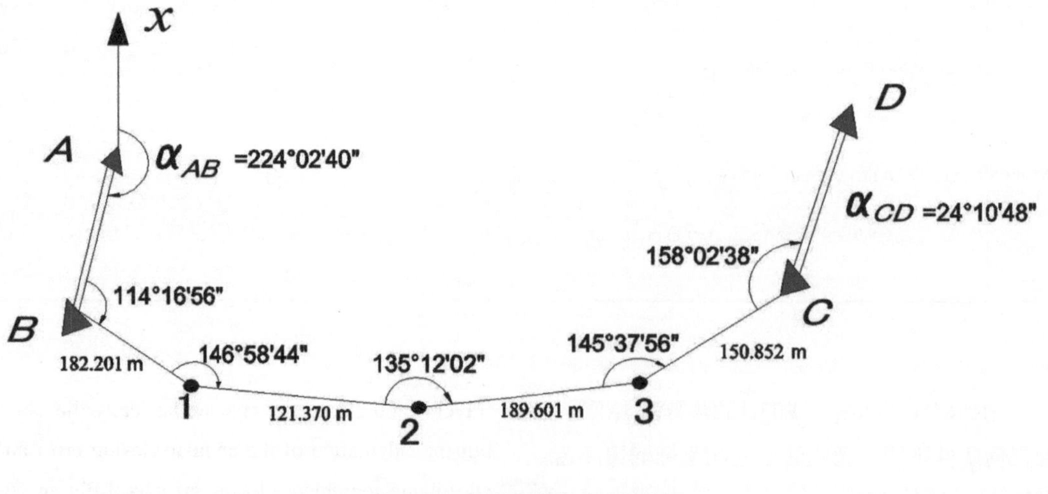


图4-10　附合导线内业计算算例

Figure 4-10 Example of Connecting Traverse Calculation

No check condition is available for the open traverse, so there is no closing error generated, and the traverse turning angle and calculated coordinate increment do not need to be corrected. The calculation steps of the open traverse are as follows.

①根据观测的转折角推算各边的坐标方位角。

① Calculate the coordinate azimuth of each leg according to the observed turning angle.

②根据各边坐标方位角和边长计算坐标增量。

② Calculate the coordinate increment according to the coordinate azimuth and length of each leg.

③根据各边的坐标增量推算各点的坐标。

③ Calculate the coordinates of each point according to the coordinate increment of each leg.

表 4-5　附合导线计算

Table 4-5 Caculation of Connecting Travese

| 点号 Point number | 观测角 Observation angle | 改正数 Correction | 改正后角值 Corrected rear angle | 坐标方位角α Goordinate azimuth α | 距离/m Distance /m | 纵坐标增量Δx Vertical coordinate increment Δx 计算值/m Calculated value /m | 改正数/mm Correction /mm | 改正后/m Corrected value /m | 横坐标增量Δy Horizontal coordinate increment Δy 计算值/m Calculated value /m | 改正数/mm Correction /mm | 改正后/m Corrected value /m | 坐标值 Coordinate value x/m | y/m |
|---|---|---|---|---|---|---|---|---|---|---|---|---|---|
| 1 | 2 | 3 | 4 | 5 | 6 | 7 | 8 | 9 | 10 | 11 | 12 | 13 | 14 |
| A | | | | | | | | | | | | | |
| | | | | 224° 02′ 40″ | | | | | | | | | |
| B | 114° 16′ 56″ | -1″ | 114° 16′ 55″ | | | | | | | | | 741.970 | 1169.386 |
| | | | | 158° 19′ 35″ | 182.201 | -169.320 | +3 | -169.317 | +67.290 | +2 | +67.292 | | |
| 1 | 146° 58′ 44″ | -2″ | 146° 58′ 42″ | | | | | | | | | 572.653 | 1236.678 |
| | | | | 125° 18′ 17″ | 121.370 | -70.143 | +2 | -70.141 | +99.049 | +3 | +99.052 | | |
| 2 | 135° 12′ 02″ | -2″ | 135° 12′ 00″ | | | | | | | | | 502.512 | 1335.730 |
| | | | | 80° 30′ 17″ | 189.601 | +31.278 | +3 | +31.281 | +187.003 | +2 | +187.005 | | |
| 3 | 145° 37′ 56″ | -2″ | 145° 37′ 54″ | | | | | | | | | 533.793 | 1522.735 |
| | | | | 46° 08′ 11″ | 150.852 | +104.532 | +3 | +104.535 | +108.763 | +2 | +108.765 | | |
| C | 158° 02′ 38″ | -1″ | 158° 02′ 37″ | | | | | | | | | 638.328 | 1631.500 |
| | | | | 24° 10′ 48″ | | | | | | | | | |
| D | | | | | | | | | | | | | |
| Σ | 700° 08′ 16″ | -8″ | 700° 08′ 08″ | | 644.024 | -103.653 | +11 | +103.642 | +462.105 | +9 | -462.114 | | |

续表

| 点号<br>Point number | 观测角<br>Observation angle | 改正数<br>Correction | 改正后角值<br>Corrected rear angle | 坐标方位角$\alpha$<br>Goordinate azimuth $\alpha$ | 距离/m<br>Distance /m | 纵坐标增量$\Delta x$<br>Vertical coordinate increment $\Delta x$ | | | 横坐标增量$\Delta y$<br>Horizontal coordinate increment $\Delta y$ | | | 坐标值<br>Coordinate value | |
|---|---|---|---|---|---|---|---|---|---|---|---|---|---|
| | | | | | | 计算值/m<br>Calculated value /m | 改正数/mm<br>Correction /mm | 改正后/m<br>Corrected value /m | 计算值/m<br>Calculated value /m | 改正数/mm<br>Correction /mm | 改正后/m<br>Corrected value /m | $x$/m | $y$/m |
| 1 | 2 | 3 | 4 | 5 | 6 | 7 | 8 | 9 | 10 | 11 | 12 | 13 | 14 |
| 辅助计算<br>Auxiliary calculation | $\Sigma\beta_{测}=700°\ 08'\ 16''$ $f_\beta=\Sigma\beta_{测}-\Sigma\beta_{理}=+8''$ $f_{允}=\pm 60\sqrt{N}=\pm 134''$ $f_\beta \leqslant f_{允}$<br>$f_x=\Sigma\Delta x_{测}-(x_c-x_b)=-0.011$ $f_y=\Sigma\Delta y_{测}-(y_c-y_b)=-0.009$<br>$f_D=\sqrt{f_x^2+f_y^2}=0.014$ $K=\frac{f_D}{\Sigma_D}\approx\frac{1}{46\ 001}$ $K_{允}=\frac{1}{15\ 000}$ $K<K_{允}$<br>$\Sigma\beta_{Observed}=700°\ 08'\ 16''$ ; $\Sigma\beta_{Observed}=700°\ 08'\ 08''$ $f_\beta=\Sigma\beta_{Observed}-\Sigma\beta_{Theoretical}=+8''$ $f_{Allowed}=\pm 60\sqrt{N}=\pm 134''$ $f_\beta \leqslant f_{Allowed}$<br>$f_x=\Sigma\Delta x_{Observed}-(x_c-x_b)=-0.011$ $f_y=\Sigma\Delta y_{Observed}-(y_c-y_b)=-0.009$<br>$K_{Allowed}=\frac{1}{15\ 000}$ $K<K_{Allowed}$ | | | | | | | | | | | | |

# 4.5 高程控制测量

# 4.5 Vertical Control Survey

## 4.5.1 高程控制网概述

## 4.5.1 Overview of Vertical Control Survey

### （1）高程控制测量概述

### (1) Overview of vertical control survey

精确测定控制点高程的工作称为高程控制测量。高程控制测量首先要在测区建立高程控制网。为了测绘地形图或建筑物施工放样以及进行科学研究工作的需要，我国在全国范围内建立一个统一的高程控制网。高程控制网由一系列的水准点构成，沿水准路线按一定的距离埋设固定的标志称为水准点。水准点分为临时性和永久性水准点，等级水准点埋设永久性标志，三、四等水准点埋设普通标石，普通水准点如图 4-11 所示。图根水准点可根据需要设置永久性或临时性水准点，临时性水准点埋设木桩或在水泥板或石头上用红油漆画出临时标志表示。

It is called the vertical control survey to accurately measure and determine the elevations of control points. For vertical control survey, a vertical control network shall be first established in the survey area. In order to facilitate the surveying and mapping of topographic maps or setting-out of buildings and the scientific research, a unified national vertical control network is established in China. The vertical control network consists of a series of bench marks, which are the fixed marks buried along the leveling line at a certain interval. Bench marks are divided into temporary and permanent ones. The graded bench marks are embedded with permanent marks. The third and fourth-order bench marks are embedded with ordinary markstones, and the ordinary bench marks are shown in Figure 4-11. The mapping bench marks can be set as permanent or temporary ones as required, and the temporary bench marks should be embedded with wooden piles or marked with red paint on cement slabs or stones.

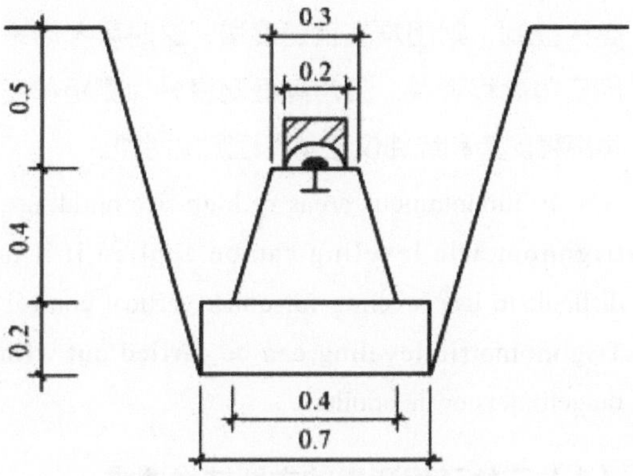


图4-11　混凝土水准标石(单位:m)

Figure 4-11 Concrete Leveling Markstone (Unit: m)

## （2）国家高程控制网

(2) National vertical control network

国家高程控制网分为一、二、三、四等四个等级。一、二等水准网是国家高程控制的基础，一、二等水准路线一般沿着铁路、公路进行布设，形成闭合水准网或附合水准网，用精密水准测量方法测定其高程；三、四等水准测量主要用于一、二等水准测量的加密，作为地形测量和工程测量的高程控制，布设闭合水准路线和附合水准路线。水准网如图 4-12 所示，国家一、二等水准测量技术指标在此不展开讲述。

The national vertical control network is divided into four orders: first-order, second-order, third-order and fourth-order. The first-order and second-order leveling networks are the basis for the national vertical control. The first-order and second-order leveling lines are generally arranged along the railways and highways to form a closed leveling network or an annexed leveling network, and their elevations are measured by precision leveling method; the third-order and fourth-order leveling is mainly used for the densification of the first and second order leveling, which is used as the vertical control of topographic survey and engineering survey, and the closed leveling line and annexed leveling route line arranged. The leveling network is shown in Figure 4-12, and the technical indicators of national first-order and second-order leveling are omitted here.

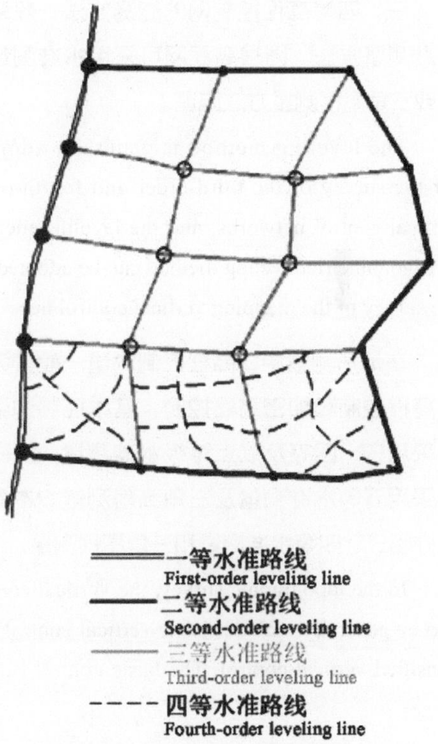


图4-12　水准网布设示意

Figure 4-12  Layout of Leveling Network

### （3）图根高程控制网

### (3) Mapping vertical control network

为了满足测图的需要，测定图根点的高程称为图根控制测量，图根高程控制网可以布设为水准网及三角高程网。

In order to meet the needs of mapping, measuring and determining the elevation of mapping control points is called mapping control survey, and the mapping vertical control network can be arranged as a leveling network and a trigonometric elevation network.

### （4）高程控制网的施测方法

### (4) Survey method of vertical control network

三、四等高程控制网的施测方法一般采用水准测量方法，图根高程网可采用水准测量方法或三角高程测量方法施测。

The leveling method is generally adopted for the survey of the third-order and fourth-order vertical control networks, and the leveling method or trigonometric leveling method can be adopted for the survey of the mapping vertical control network.

在地形测量中，高程控制测量一般分为基本高程控制和加密高程控制，基本控制测量一般采用三、四等及以上等级水准测量，加密一般采用五等水准测量及三角高程测量。本章主要介绍三、四等水准测量和三角高程测量。

In the topographic survey, the vertical control survey generally includes basic vertical control and densified vertical control. The basic control survey generally adopts third-order, fourth-order and above leveling, and the densified survey generally adopts fifth-order leveling and trigonometric leveling. This chapter mainly introduces the third-order and fourth-order leveling and trigonometric leveling.

## 4.5.2 三角高程测量

## 4.5.2 Trigonometric Leveling

在山区或高层建筑物上，若用水准测量作高程控制，则困难大且速度慢，这时可考虑采用三角高程测量。三角高程测量分为测距仪三角高程测量和经纬仪三角高程测量两种。

In mountainous areas or high-rise buildings, trigonometric leveling can be applied if it is difficult to use leveling for quick vertical control. Trigonometric leveling can be carried out with rangefinders or theodolites.

### （1）三角高程测量的主要技术要求

### (1) Main technical requirements for trigonometric leveling

三角高程测量的主要技术要求是针对竖直角测量的技术要求，一般分为两个等级，即四等、五等。三角高程测量可作为测区的首级控制，技术要求见表 4-6。

The main technical requirements of trigonometric leveling are the technical requirements for vertical angle measurement, which are generally divided into two grades, i.e. fourth-order and fifth-order. Trigonometric leveling can be used as the primary control of the survey area, its technical requirements are shown in Table 4-6.

表 4-6 三角高程测量的主要技术要求

Table 4-6 Main Technical Requirements for Trigonometric Leveling

| 等级 Class | 仪器 Instru-ment | 测距边 Ranging edge 测回数 Number of observation sets | 竖直角测回数 Number of observation sets for vertical angle measurement | | 指标差较差/(″) Difference between index errors/(″) | 竖直角较差/(″) Difference between observation sets/(″) | 对向观测高差较差/mm Discrepancy between elevation differences measured in reciprocal observations/mm | 附合或环线闭合差/mm Closing error of annexed leveling line or closed leveling line/mm |
|---|---|---|---|---|---|---|---|---|
| | | | 三丝法 Three-wire leveling method | 中丝法 Central-wire leveling method | | | | |
| 四等 Fourth-order | $DJ_2$ | 往返各一次 Round trip once | — | 3 | ≤7 | ≤7 | $40\sqrt{D}$ | $20\sqrt{\Sigma_D}$ |
| 五等 Fifth-order | $DJ_2$ | 1 | 1 | 2 | ≤10 | ≤10 | $60\sqrt{D}$ | $30\sqrt{\Sigma_D}$ |
| 注：$D$为电磁波测距边长度(km)。 Note: $D$ is the length of electromagnetic ranging edge (km). | | | | | | | | |

## （2）三角高程测量原理

### (2) Principle of trigonometric leveling

三角高程测量是根据两点间的水平距离和竖直角计算两点的高差，然后求出所求点的高程。

Trigonometric leveling is to calculate the elevation difference between two points according to the horizontal distance and vertical angle between two points, and then calculate the elevation of the obtained point.

如图 4-13 所示，在 $A$ 点安置仪器，用望远镜中丝瞄准 $B$ 点觇标的顶点，测得竖直角 $\alpha$ 觇并量取仪器高 $i$ 和觇标高 $s$，若测出 $A$、$B$ 两点间的水平距离 $D$，则可求得 $A$、$B$ 两点之间的高差，即

As shown in Figure 4-13, place the instrument at point $A$, aim at the vertex of the survey target at point $B$ with the telescope, measure the vertical angle $\alpha$, and measure the instrument elevation $i$ and the survey target elevation $s$. If the horizontal distance $D$ between points $A$ and $B$ is measured, the elevation difference between points $A$ and $B$ can be obtained, i.e.

$$h_{AB}=D\tan\alpha+i-s \tag{4.18}$$

$B$ 点高程为

The elevation of $B$ can be calculated by the following formula

$$H_B=H_A+D\tan\alpha+i-s \tag{4.19}$$

三角高程测量一般应采用对向观测法。如图 4-13 所示，即由 $A$ 向 $B$ 观测称为直觇，再由 $B$ 向 $A$ 观测称为反觇，直觇和反觇称为对向

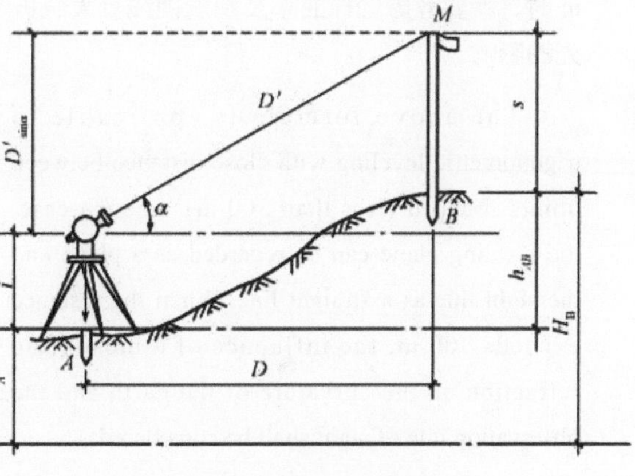


图4-13 三角高程测量原理

Figure 4-13 Principle of Trigonometric Leveling

观测。采用对向观测的方法可以减弱地球曲率和大气折光的影响。对向观测所求得的高差较差不应大于 $0.1D$（$D$ 为水平距离，以 km 为单位，其结果以 m 为单位）。取对向观测的高差中数为最后结果，即

Reciprocal observation method is generally adopted for trigonometric leveling. As shown in Figure 4-13, the observation from $A$ to $B$ is called forward survey, and the observation from $B$ to $A$ is called backward survey, and forward survey and backward survey is called reciprocal observation. The application of reciprocal observation can reduce the effects of earth curvature and atmospheric refraction. The discrepancy between the elevation differences obtained by reciprocal observation shall not be greater than $0.1D$ ($D$ is the horizontal distance, in km, and the result is in m). Take the median of the elevation difference in the reciprocal observation as the final result, i.e.

$$h_{中} = \frac{1}{2}(h_{AB} - h_{BA}) \tag{4.20}$$

$$h_{\text{Median}} = \frac{1}{2}(h_{AB} - h_{BA}) \tag{4.20}$$

上述公式适用于 $A$、$B$ 两点距离较近（小于 300 m）的三角高程测量，此时水准面可近似看成平面，视线视为直线。当距离超过 300 m 时，就要考虑地球曲率及观测视线受大气折光的影响。

The above formula is applicable to trigonometric leveling with close distance between points $A$ and $B$ (less than 300 m). In such case, the leveling plane can be regarded as a plane and the sight line as a straight line. When the distance exceeds 300 m, the influence of atmospheric refraction on the curvature of the earth and the observation line of sight shall be considered.

### （3）三角高程测量的观测和计算

### (3) Observation and calculation of trigonometric leveling

三角高程测量的观测与计算应按以下步骤进行。

The observation and calculation of trigonometric leveling shall be carried out according to the following steps.

①安置仪器于测站上，量出仪器高 $i$，觇标立于测点上，量出觇标高 $s$。

① Place the instrument on the observation station, measure the elevation of the instrument $i$, erect a survey target on the survey point, and measure the elevation of the survey target $s$.

②用经纬仪或测距仪采用测回法观测竖直角 $\alpha$，取其平均值为最后观测结果。

② Observe the vertical angle $\alpha$ with a theodolite or a rangefinder and take the average value as the final observation result.

③采用对向观测，其方法同前两步。

③ Apply the reciprocal observation method and take the previous two steps.

④用式（4.19）和式（4.20）计算高差和高程。

④ Calculate the elevation difference and elevation by formula (4.19) and formula (4.20).

三角高程路线，尽可能组成闭合测量路线或附合测量路线，并尽可能起闭于高一等级的水准点上。若闭合差在所规定的容许范围内，则将 $f_h$ 反符号按照与各边边长成正比例的关系分配到各段高差中，最后根据起始点的高程和改正后的高差计算出各待求点的高程。

The trigonometric leveling line shall form a closed leveling line or an annexed leveling line as far as possible, and shall be started and closed on the bench mark of a higher order as far as possible. If the closing error $f_h$ is within the allowable range, the $f_h$ is evenly distributed to each elevation difference with the opposite sign in direct proportion to the side length, and finally the elevation of each point is calculated according to the elevation of the starting point and the corrected elevation difference.

## 4.5.3 GPS 拟合高程测量

## 4.5.3 GPS Leveling Fitting

GPS 拟合高程测量仅适用于平原或丘陵地区的五等及五等以下等级高程测量。GPS 拟合高程测量宜与 GPS 平面控制测量一起进行。

GPS leveling fitting is only applicable to fifth-order and below leveling in plain or hilly areas. GPS leveling fitting should be carried out together with GPS horizontal control survey.

### （1）GPS 拟合高程测量的主要技术要求

### (1) Main technical requirements for GPS leveling fitting

① GPS 网应与四等或四等以上的水准点联测。联测的 GPS 点，宜分布在测区四周和中央。若测区为带状地形，则联测的 GPS 点应分布于测区两端及中部。

① The GPS network shall be connected with the fourth-order or above leveling points for connection survey. The GPS points for connection survey should be distributed around and in the center of the survey area. If the survey area is of strip-shaped terrain, the GPS points for connection survey shall be distributed at both ends and in the middle of the survey area.

②联测点数宜大于选用计算模型中未知参数个数的 1.5 倍，点间距宜小于 10 km。

② The number of points for connection survey should be greater than 1.5 times the number of unknown parameters in the selected calculation model, and the point spacing should be less than 10 km.

③地形高差变化较大的地区，应适当增加联测的点数。

③ In areas with large changes in terrain elevation difference, the number of points for connection survey shall be appropriately increased.

④地形趋势变化明显的大面积测区，宜采取分区拟合的方法。

④ The method of subregional fitting should be adopted for large area survey areas with obvious topographic trend changes.

⑤ GPS 观测的技术要求，应按有关规定执行；其天线高应在观测前后各测量一次，取其平均值作为最终高度。

⑤ The technical requirements for GPS observation shall be implemented according to relevant regulations. The antenna height shall be measured once respectively before and after the observation, and the average value shall be taken as the final height.

### （2）GPS 拟合高程计算

### (2) Elevation calculation for GPS leveling fitting

GPS 拟合高程计算应充分利用当地的重力大地水准面模型或资料，应对联测的已知高程点进行可靠性检验，并剔除不合格点。对于地形平坦的小测区，可采用平面拟合模型；对于地形起伏较大的大面积测区，宜采用曲面拟合模型。对拟合高程模型应进行优化。

For the elevation calculation for GPS leveling fitting, the local gravity geoid model or data shall be fully utilized, the known elevation points of connection survey shall be checked for reliability, and the unqualified points shall be eliminated. For small survey areas with flat terrain, plane fitting model can be adopted; for large survey areas with large topographic relief, curved surface fitting model should be adopted. The elevation fitting model shall be optimized.

GPS 点的高程计算，不宜超出拟合高程模型所覆盖的范围。

For the elevation calculation of GPS points, the range covered by the elevation fitting model should not be exceeded.

### （3）GPS 点的拟合高程成果检验

### (3) Results check for the elevation calculation of GPS points by the elevation fitting model

检测点数不少于全部高程点的 10% 且不少于三个点；高差检验可采用相应等级的水准测量方法或电磁波测距三角高程测量方法进行，其高差较差不应大于 30 $\sqrt{D}$ mm（$D$ 为检查路线的长度，单位为 km）。

The number of check points shall not be less than 10% of all elevation points and shall not be less than 3 points. The elevation difference check can be carried out by the leveling method or electromagnetic distance measurement trigonometric leveling method of corresponding orders, and the discrepancy between elevation differences shall not be greater than $30\sqrt{D}$ mm ($D$ is the length of the checked line, in km).

# 本章小结

## Chapter Summary

本章主要讲述了国家平面和高程控制网、导线测量、高程控制测量等内容。

This chapter mainly describes the national horizontal and vertical control network, traverse survey, vertical control survey, etc.

1. 在测区内，按测量任务所要求的精度，测定一系列控制点的平面位置和高程，建立起测量控制网，作为各种测量的基础，这种测量工作称为控制测量。测定控制点平面位置的工作称为平面控制测量，测定控制点高程的工作称为高程控制测量。

1.In the survey area, it is called the control survey to measure and determine the plane positions and elevations of a series of control points according to the accuracy required by the survey task and then to establish a survey control network as the basis for various survey tasks. It is called the horizontal control survey to measure and determine the plane positions of control points. It is called the vertical control survey to measure and determine the elevations of control points.

2. 平面控制测量的主要方法有三角测量和导线测量。

2.The main methods of horizontal control survey are trigonometric survey and traverse survey.

3. 将相邻控制点依次用直线相连而组成的折线称为导线，构成导线的控制点称为导线点。导线测量就是依次测量各导线边的水平距离以及相邻导线边的水平夹角，然后根据起算数据，推算各导线边的坐标方位角，从而求出各导线点的平面坐标。

3.The polyline formed by connecting adjacent control points in turn with straight lines is called a traverse, and the control points that make up the traverse are called traverse points. The traverse survey is to measure the horizontal distance between traverse legs and the horizontal angles between adjacent traverse legs in turn, and then calculate the coordinate azimuth angles of various traverse legs according to the known data, so as to obtain the horizontal coordinates of various traverse points.

4. 高程控制测量的方法主要有水准测量和三角高程测量。高程控制测量精度等级的划分，依次为二等、三等、四等、五等。各等级高程控制宜采用水准测量，四等及以下等级可采用电磁波测距三角高程测量，五等也可采用 GPS 拟合高程测量。

4.Vertical control survey methods mainly include leveling and trigonometric leveling. There are generally four orders of vertical control survey accuracy, namely, second-order, third-order, fourth-order and fifth-order. Leveling should be adopted for each order of vertical control, generally trigonometric leveling with electromagnetic distance measurement for fourth-order or below, and GPS leveling fitting for fifth-order.

# 复习思考题

# Review Questions

## 一、填空题

## I. Fill in the Blanks

1. 测定控制点平面位置的工作称为______________；测定控制点高程的工作称为______________。

1. The work of measuring and determining the plane position of control points is called______________; the work of measuring and determining the elevation of control points is called______________.

2. 在全国范围内统一建立的三角网，称为______________。

2. The triangular network established nationwide is called______________.

3. 各等级高程控制宜采用______________，四等及以下等级可采用______________，五等也可采用______________。

3. Each order of vertical control should be carried out by the method of ______________; generally, fourth-order leveling or below can be performed by the method of______________, and fifth-order leveling can be conducted by the method of______________.

4. 在小区域（面积______________）内建立的平面控制网，称为小区域平面控制网。

4. The horizontal control network established in a small area (area______________) is called the small area horizontal control network.

5. 根据测区的不同情况和具体要求，导线可按____________、____________、____________三种形式进行布设。

5. According to different conditions and specific requirements of the survey area, the traverse can be laid in the following three forms, namely, _____________, ____________, ____________ .

6. 导线测量的外业工作包括____________、____________、__________和____________。

6. The field work of traverse survey includes__________,__________,__________and____________.

## 二、选择题

## II. Choice Questions

1. 图 4-14 所示的导线属于（ ）。

1.The traverse shown in Figure 4-14 belongs to ( ).

A. 闭合导线

A. closed traverse

B. 附合导线

B. connecting traverse

C. 支导线

C. open traverse

D. 以上答案都正确

D. all above

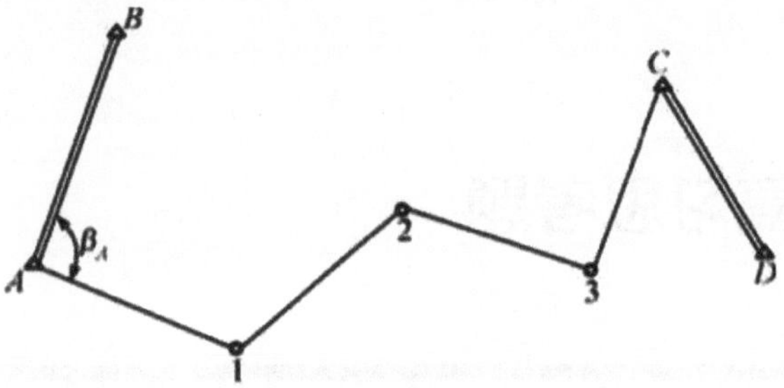


图4-14 导线示意

Figure 4-14 Schematic Diagram of Traverse

2. 导线测量角度闭合差的调整方法是（ ）。

2.The adjustment method for closing error of traverse survey is to ( ).

A. 反号按角度个数平均分配

A. evenly distribute the closing error to each angle with the opposite sign

B. 反号按角度大小比例分配

B. distribute the closing error to each angle with the opposite sign in proportion to the angle size

C. 反号按边数平均分配

C. evenly distribute the closing error with the opposite sign according to the number of legs

D. 反号按边长比例分配

D. distribute the closing error with the opposite sign in proportion to the leg length

3. 衡量导线测量精度的一个重要指标是（ ）。

3. An important indicator to measure the accuracy of traverse is ( ).

A. 坐标增量闭合差

A. increment closing error of coordinate

B. 导线全长闭合差

B. total length closing error of traverse

C. 导线全长相对闭合差

C. relative total length closing error of traverse

D. 以上答案都正确

D. all above

4. 首级控制网定向时，方位角传递宜联测（　　）个方向?

4. When the primary control network is oriented, how many directions should be selected for connection survey for the azimuth transfer ?

A. 1　　　　B. 2　　　　C.3　　　　D.4

5. 三等、四等水准测量一般采用（　　）观测。

5. Which method is generally used for third-order and fourth-order leveling ?

A. 单面尺法　　　　B. 双面尺法

A. one-sided leveling rod method　　　　B. double-sided leveling rod method

C. 方向观测法　　　　D. 回测法

C. direction observation method　　　　D. forward and backward observation

## 三、简答题

## III. Short Answer Questions

1. 什么是控制测量？为什么要进行控制测量?

1.What is control survey? Why do we carry out the control survey?

2. 导线的布设形式有哪些？试绘图说明。

2.What are the layout forms of traverse? Please illustrate with drawings.

3. 简述导线测量的外业工作。

3.Please give a brief description of the field work of traverse survey.

4. 进行三等、四等水准测量时，测站的观测程序是什么?

4.What is the observation sequence for an observation station during the third-order and fourth-order leveling?

5. 在什么情况下采用三角高程测量？为什么要采用对向观测?

5.When do we need to use trigonometric leveling? Why should we adopt reciprocal observation?

5

# 地形图的测绘与应用

# Surveying, Mapping and Application of Topographic Map

**学习目标：**

**Learning Objectives:**

通过本章的学习，了解地形图的概念、地形图测绘前的准备工作；熟悉比例尺及其精度、地形图绘制的基本要求、地形图的识读方法；掌握大比例尺地形图的分幅与编号方法，地物、地貌在地形图上的表示方法及等高线的特征，地形图的测绘方法，地形图在工程建设中的具体应用。

Through the study of this chapiter, to learn about the concept of a topographic map and the preparations before surveying and mapping a topographic map; to get familiar with the scale and scale accuracy, the basic requirements for surveying and mapping a topographic map, and the reading method of a topographic map; to master the subdivision and numbering methods of large-scale topographic maps, the representation methods of ground objects and land features on topographic maps, the features of contour lines, the surveying and mapping methods of topographic maps, and the specific application of topographic maps in engineering construction.

**能力目标：**

**Ability Objectives:**

能够正确阅读、绘制地形图，具备测绘大比例尺地形图的基本技能，能够利用地形图确定最佳线路方案，能够完成断面图的绘制，确定汇水面积并根据地形图进行场地平整。

Be able to correctly read and draw topographic maps, have the basic skills of surveying and mapping large-scale topographic maps, and be able to select the optimal route scheme, draw cross-section diagrams, figure out the catchment area and level the site based on topographic maps.

# 5.1 地形图的基本知识

## 5.1 Basic Knowledge of Topographic Map

### 5.1.1 地形图的概念

### 5.1.1 Concept of Topographic Map

地形图是通过实地测量，将地面上各种地物、地貌的平面位置和高程位置，按一定的比例尺，用统一规定的符号和注记缩绘在图纸上的平面图形。它既表示地物的平面位置，又表示地貌形态。地物是指地球表面上轮廓明显、具有固定性的物体。地物又分为人工地物（如道路、房屋等）和自然地物（如江河、湖泊等）。地貌是指地球表面高低起伏的形态（如高山、丘陵、平原、洼地等）。地物和地貌统称为地形。

A topographic map is a plane figure bearing plane positions and elevation positions of ground objects and land features on the ground that are scaled down at a scale and represented with uniform symbols and notes. It shows both the plane positions of ground objects and the land features. A ground object is a fixed object that has a clear contour on the earth's surface. Ground objects can be categorized as artificial ground objects (such as roads and houses) and natural ground objects (such as rivers and lakes). A land feature refers to the undulating shape of the earth's surface (such as mountains, hills, plains, and depressions). Ground objects and land features are collectively referred to as terrain.

地形图是地球表面实际情况的客观反映，各项经济和国防建设工程都需首先在地形图上进行规划、设计，特别是大比例尺（常用的有 1 ∶ 500、1 ∶ 1 000、1 ∶ 2 000、1 ∶ 5 000 等）地形图，也是城乡建设和各项建筑工程进行规划、设计、施工的重要基础资料之一。

A topographic map is a reflection of the objective realities of the earth's surface, and a crucial basis for planning and design of construction works for economic development and national defense development. Especially, a large-scale topographic map (commonly 1 ∶ 500, 1 ∶ 1 000, 1 ∶ 2 000, 1 ∶ 5 000, etc.) presents fundamental information for the planning, design and execution of construction works for urban and rural development and architectural projects.

### 5.1.2 地形图的比例尺

### 5.1.2 Scale of Topographic Map

地形在图上的线段长度和地面相应长度之比称为地形图比例尺。

The scale of a topographic map refers to the ratio of the line dimension of the terrain on the map to its corresponding dimension on the ground.

#### (1) 数字比例尺

#### (1) Numeric scale

一般用分子为 1 的分数表示的比例尺称为数字比例尺。设图上的某一直线长度为 $d$，地面上相应线段的水平距离为 $D$，则

Numeric scale is indicated by a fraction whose numerator is usually 1. Here we assume that the length of a straight line on the map is $d$ and the corresponding horizontal distance on the ground is $D$. Then, we can get the formula as follows

$$\frac{d}{D}=\frac{1}{M} \tag{5.1}$$

式中 $M$ 是地形图比例尺分母；

Where, $M$ is denominator of the scale;

$d$ 是地形图上某线段的长度；

$d$ is length of a line segment on the map;

$D$ 是实地相应的投影长度。

$D$ is corresponding projection length on the ground.

公式可用于地形图上的线段与实地对应线段投影长度之间的换算；同理，还可以求出地图上某区域面积与实地对应区域的投影面积之比的关系式，即

The formula can be used for conversion between the length of the line segment on the map and its projection length on the ground. Similarly, the relation between the area of a certain range on the map and the projection area of the corresponding range on the ground can also be derived from this formula, as follows.

$$\frac{1}{m^2}=\frac{f}{F} \tag{5.2}$$

式中 $f$ 是地形图上某区域的面积；

Where, $f$ is area of a certain range on the map;

$F$ 是实地对应区域的投影面积。

$F$ is projected area of the corresponding range on the ground.

要把地面上的线段描绘到地形图平面上，首先将地面线段沿垂线投影到大地水准面上，然后归化到椭球体面上，再按某种方法将其投影到平面上，最后按某一比率将它缩小到地形图上，这个缩小比率就是地形图比例尺。在工程建设中，通常使用的比例尺有 1 ∶ 500、1 ∶ 1 000、1 ∶ 2 000、1 ∶ 5 000、1 ∶ 10 000、1 ∶ 25 000、1 ∶ 50 000 等。用分数形式表示的比例尺称为数字比例尺，分数值越大，比例尺也越大，图上表示的地物、地貌也越详细。

To draw a line segment on a map to indicate a real line on the ground, we need to first project the line along a vertical line to the geoid and transfer it onto an ellipsoid surface; then we project it on the plane by some means and scale it down on the map at a ratio, which is the scale of the topographic map. Scales such as 1 ∶ 500, 1 ∶ 1,000, 1 ∶ 2 000, 1 ∶ 5 000, 1 ∶ 10 000, 1 ∶ 25 000 and 1 ∶ 50 000 are commonly used in engineering construction. The scale expressed as a fraction is called a numeric scale. The larger the fraction, the larger the scale, and the more detailed the ground objects and land features represented on the map.

### (2) 直线比例尺

### (2) Bar scale

在图上绘制一条线段，将其分成若干段，并将其代表的实地长度标注上，可将图上量取的长度与之比较得出实地距离，这种比例尺称为直线比例尺。如图 5-1 所示，在直线上截取若干基本单位（如 2 cm），将左端的基本单位再 10 等分（如 2 mm）。对于某种比例尺，如 1 ∶ 10 000 比例尺，直线上每 2 cm 及 2 mm 分别相当于地面 200 m 及 20 m。利用直线比例尺，可以量取图上两点间所代表的实地距离。

A bar scale is a line divided into several segments, labeled with ground lengths, and can be used to find the real-world distance as compared to the length of the segment on the map. As shown in Figure 5-1, a line is divided into several basic units (e.g., 2 cm) and then a basic unit is subdivided into 10 equal parts (e.g., 2 mm) as shown at the left end. For a scale, such as 1 ∶ 10 000, every 2 cm and 2 mm on the line is equivalent to 200 m and 20 m on the ground, respectively. The real-world distance represented by a segment between two points on the map can be measured with the bar scale.

制作方法：画两条平行于图纸底边的平行线，将直线以 2 cm 为基本单位进行等分，将最左端的 2 cm 再 10 等分，并定出零点。将 2 mm

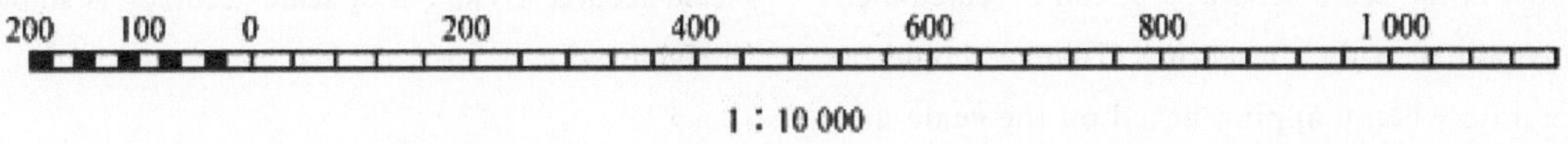


图5-1　直线比例尺

Figure 5-1 Bar Scale

及 2 cm 所表示的实地水平距离，依等分倍数标到图上相应的位置。

How to draw a bar scale: Draw two straight lines parallel to the bottom edge of the drawing paper, divide the line into equal parts with 2 cm as a basic unit, subdivide the 2 cm into 10 equal parts at the left end, and give a zero point. Mark the real-world horizontal distances represented by 2 mm and 2 cm segments to the corresponding positions by multiples based on the division distance.

使用方法：先用两角规对准图上待测线段的两端点，再拿到比例尺图上比测（角规一脚对准 2 cm的分化线，另一脚须位于毫米划分区间）。

How to use a bar scale: Put two legs of a divider on the two ends of the line segment to be measured on the drawing, and then put the divider on the bar scale for comparison and measurement (with one leg of the divider on the 2 cm division line and the other leg on the 2 mm division section).

优点：直线比例尺与数字比例尺相比，具有表现手段直观，使用方便，并能减少由于图纸伸缩而引起的距离换算误差等优点。

Advantages: Compared with a numeric scale, the bar scale provides a visual indication of distance, can be conveniently used and reduce errors of distance conversion after scaling.

原因：地形图长期存放，由于受潮或干燥产生伸缩变形，此时如仍使用原数字比例尺计算与其对应的水平距离，就会产生误差。如使用直线比例尺，因其附在地形图上，与地形图同步伸缩，故避免了由图纸伸缩变形而引起的图上距离量测误差。

Reason: When a topographic map is left unused for a long time, humidity may result in distortion and expansion of map paper, and errors may occur if the original numeric scale is used to calculate the corresponding horizontal distance. While the bar scale is marked on the topographic map and will also be distorted as the distortion of the map, thus avoiding the distance measurement error caused by the map distortion.

## 5.1.3 比例尺精度

### 5.1.3 Scale Accuracy

地形图所能达到的最大精度取决于人眼的分辨能力和绘图与印刷的能力。其中，人眼的分辨能力是主要的因素。对人眼的分辨能力的分析可知，在一般情况下，人眼的最小鉴别角为 $\theta$=60"。若以明视距离 250 mm 计算，则人眼能分辨出的两点间的最小距离约为 0.1 mm，它在地上表示的水平距离为 0.1 mm × $M$，称为比例尺精度。利用比例尺精度，根据比例尺可以推算出测图时量距应准确到什么程度。比例尺的精度见表 5-1。

The maximum accuracy of a topographic map depends on the resolution ability of human eyes and the quality of mapping and printing. Among them, the ability of human eyes to distinguish is the main factor. The minimum angle of resolution of a human eye is usually $\theta$=60" based on the analysis of the resolution of human eyes. If the distance of distinct vision is 250 mm, the minimum distance between two points that can be resolved by human eyes is about 0.1 mm. Therefore, the corresponding horizontal distance on the ground is 0.1 mm×$M$, which is the scale accuracy. It can be calculated to what extent the measuring distance should be accurate when mapping based on the scale and scale accuracy. The list of scale accuracy is shown in Table 5-1.

表 5-1　比例尺的精度

Table 5-1 Scale Accuracy

| 比例尺<br>Scale | 1 : 500 | 1 : 1 000 | 1 : 2 000 | 1 : 5 000 | 1 : 10 000 |
|---|---|---|---|---|---|
| 比例尺的精度/m<br>Accuracy/m | 0. 05 | 0. 1 | 0. 2 | 0. 5 | 1. 0 |

例如，1 ： 1 000 地形图的比例尺精度为 0.1 m，测图时量距的精度只需 0.1 m，小于 0.1 m 的距离在图上表示不出来；反之，根据图上表示实地的最短长度，可以推算测图比例尺。例如，表示实地最短线段长度为 0.5 m，则测图比例尺不得小于 1 ： 5 000。

For example, the scale accuracy of a 1 : 1 000 topographic map is 0.1 m, so the accuracy of measuring distance is only accurate to 0.1 m. A distance less than 0.1 m cannot be represented on the map; on the contrary, the mapping scale can be calculated according to the shortest length that can be resolved on a map to show the real-world distance. For example, if the mentioned shortest length is 0.5 m, the mapping scale shall not be smaller than 1 : 5 000.

比例尺越大，采集的数据信息越详细，精度要求就越高，测图工作量和投资往往成倍增加。因此，使用何种比例尺测图应从实际需要出发，不应盲目追求更大比例尺的地形图。

The larger the scale, the more detailed the collected data information, the larger the required accuracy, and the mapping workload and investment are often multiplied. For this concern, the type of scale must be chosen based on actual needs, and a larger scale is not always the best choice.

【例 5.1】甲方要求测绘 1 ： 2 000 比例尺的地形图，在测绘地形图时，距离测量应精确到什么程度呢？

[Example 5.1] The client requires mapping a topographic map with a scale of 1 : 2 000. What should be the mapping accuracy of this map?

【解】根据题意，应精确到

[Solution] Based on the information given in the question, it should be accurate to

$$\delta = 0.1 \times 2\,000 = 200 \text{ mm} = 0.2 \text{ m}$$

## 5.1.4 地形图的分幅、编号和图廓

## 5.1.4 Subdivision, Numbering and Neatline of Topographic Map

### (1) 地形图分幅

### (1) Subdivision of topographic map

当测区范围较大时，由于图纸尺寸有限，不可能将测区内的所有地形都绘制在一幅图内。这时，我们需要分幅绘制地形图。为了便于测绘、使用和保管地形图，常将地形图按照一定的规则进行分幅和编号。

When the surveying area is large, it is impossible to map all the features and land features in the surveying area in one map due to the limited size of the drawing. Therefore, it is necessary to subdivide a topographic map in different sheets. Topographic maps are often subdivided and numbered based on some rules to facilitate the mapping, use and storage.

## （2）地形图编号

### (2) Numbering of topographic map

图号是为了方便储存、检索和使用地形图而给予各幅地形图的代号，通常标注在地形图的正上方处。地形图的分幅方法有两类：一类是按经纬线分幅的梯形分幅法，它一般用于 1 ： 5 000 至 1 ： 1 000 000 的中、小比例尺地形图的分幅；另一类是按坐标格网分幅的矩形分幅法，它一般用于城市和工程建设 1 ： 500 至 1 ： 2 000 的大比例尺地形图的分幅。地形图的分幅如图 5-2 所示。

A map is usually attached with a number right above the topographic map to facilitate storage, retrieval and use. There are two methods for subdividing a topographic map: One is the trapezoidal subdivision method based on longitude and latitude, which is generally used for subdividing medium-scale and small-scale topographic maps with a scale of 1 ： 5 000 to 1 ： 1 000 000; the other is the rectangular subdivision method, which is generally used for large-scale topographic maps that usually for urban and engineering construction, with a scale of 1 ： 500 to 1 ： 2 000. The subdivision of topographic maps is shown in Figure 5-2.

编号方法有以下几种：

The numbering methods are as follows:

①以 1 ： 5 000 比例尺地形图西南角坐标为基础图号 X-Y，如 20-30；下一级比例尺图幅的编号是在此基础图号的后面加 I、II、III、IV。矩形图幅分幅和编号分别见图 5-3、表 5-2。

① On a topographic map with a scale of 1 ： 5 000, the coordinates of the southwest corner are taken as the basic number X-Y, such as 20-30; the number of the subdivided map sheet with a scale of the immediate next order is added with I, II, III and IV after this basic number. Refer to Figure 5-3 and Table 5-2 for the subdivision and numbering of topographic map with rectangular format method.

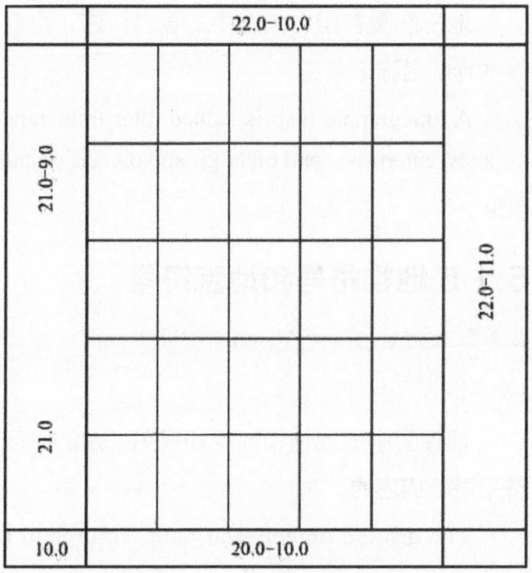


图5-2　地形图的分幅图

Figure 5-2 Subdivision of Topographic Map

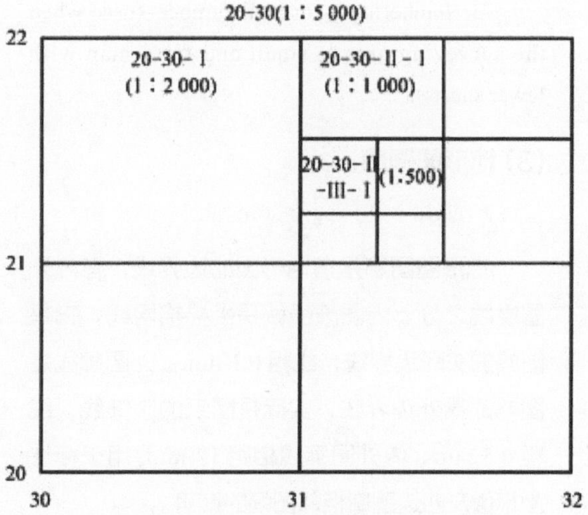


图5-3　矩形图幅分幅和编号

Figure 5-3 Subdivision and Numbering of Topographic Map with Rectangular Format Method

表 5-2 矩形图幅分幅

Table 5-2 Subdivision and Numbering of Topographic Map with Rectangular Format Method

| 比例尺 Scale | 图幅大小/cm² Sheet size/cm² | 实地面积/km² Actual area /km² | 1 : 5 000图中含该比例尺图幅数 Number of sheets with the listed scale (in the first column) in a map with a scale of 1 : 5000 | 图幅编号 Sheet No. | |
|---|---|---|---|---|---|
| | | | | ① | ② |
| 1 : 5 000 | 40 × 40 | 4 | 1 | 20-30 | 20-30 |
| 1 : 2 000 | 50 × 50 | 1 | 4 | 20-30-I | 21.0-30.0 |
| 1 : 1 000 | 50 × 50 | 0.25 | 16 | 20-30-II-I | 21.5-31.0 |
| 1 : 500 | 50 × 50 | 0.0625 | 64 | 20-30-II-III-I | 21.25-3 100 |

②采用地形图西南角坐标的千米数进行编号。1 ： 5 000 地形图取至整千米数；1 ： 2 000 地形图和 1 ： 1 000 地形图取至 0.1 km。

② The numbering is based on the coordinates of the southwest corner of the topographic map, which are numbers with a measurement unit of kilometers. The sheet number of a topographic map with a scale of 1 ： 5 000 is accurate to an integer and that with a scale of 1 ： 1 000 or 1 ： 2 000 is accurate to 0.1 km.

③自然序数法编号（测区较小，图幅数不多的情况下使用）。

③ Numbering by ordinal numbers (used when the surveying area is small and for a map with fewer sheets).

### (3) 地形图图廓

(3) Neatline of topographic map

地形图图廓是指地形图的边界线，有内外图廓线之分。内图廓线就是坐标格网线，它是图幅的实际边界线，线粗 0.1 mm。外图廓线是图幅的最外边界线，实际是图纸的装饰线，线粗 0.5 mm。内外图廓线相距 12 mm, 用于标注坐标值及四边图廓相邻图幅的编号。

Neatlines refer to the border lines of a topographic map and can be categorized into inner neatline and outer neatline. The inner neatline is the coordinate grid line, which is the actual border line of a map sheet, with a line thickness of 0.1 mm. The outer neatline is the outermost border line of the map only for decoration, with a line thickness of 0.5 mm. The distance between the inner and outer neatlines is 12 mm, and the space between them is used for marking the coordinate values and the numbers of map sheets adjacent to the neatlines on four sides.

### （4）地形图的图名

(4) Name of topographic map

地形图图名用本幅图中重要的村庄、企业等地物的名称来命名。

A topographic map is named after influential villages, enterprises and other ground objects on the map.

## 5.1.5 地物符号和地貌符号

## 5.1.5 Symbols of Ground Objects and Land Features

地形图中的地物和地貌都是用规定的图示符号进行描述的。

The ground objects and land features in a topographic map are usually depicted by specified graphic symbols.

## （1）地物符号

## (1) Symbols of ground objects

根据地物的大小和描绘方法的不同，地物符号可以分成依比例符号、非比例符号、线形符号和地物注记四种类型。

Symbols of ground objects are categorized into proportional symbols, non-proportional symbols, linear symbols and notes of ground objects based on the size and depiction methods.

①依比例符号。当地物的轮廓尺寸较大时，常按测图的比例尺将其形状大小缩绘到图纸上，绘出的符号称为依比例符号，如一般房屋、简易房屋等符号。

① Proportional symbol. For a ground object with large outline dimensions, the shape and size of the ground object are usually scaled down at the map scale and then depicted on the map. The symbol depicted on the map is called a proportional symbol, such as ordinary and makeshift houses.

②非比例符号。当地物的轮廓尺寸较小时，如三角点、水准点、独立树、消火栓等，无法将其形状和大小按测图的比例尺缩绘到图纸上，但这些地物很重要，必须在图上表示出来，则不管地物的实际尺寸大小，均用规定的符号表示在图上，这类符号称为非比例符号。

② Non-proportional symbols. For a ground object with small outline dimensions, such as trigonometrical points, benchmarks, separate trees and fire hydrants, it is difficult to scale down their shape and size at the map scale, but these ground objects are very important and must be shown on the drawings. In this case, they will be indicated on the map with specified symbols regardless of the actual size. This kind of symbol is called a non-proportional symbol.

非比例符号的中心位置的确定，随各种不同的地物而异，在测图和用图时应注意下列几点。

The central position of a non-proportional symbol varies with different ground objects. The following shall be noted when mapping and using maps.

a. 圆形、矩形、三角形等单个几何图形符号，定位点在其几何图形的中心。如三角点、水准点等。

a. For a single geometry symbol of a circle, rectangle and triangle, such as triangulation points and benchmarks, the positioning point is in its center.

b. 宽底符号（蒙古包、烟囱等），定位点在底线中心。

b. The positioning point of a wide-bottom symbol (Mongolian yurt, chimney, etc.) is in the center of the bottom line.

c. 底部为直角形的符号（风车、路标等），定位点在直角的顶点。

c. The positioning point of a symbol with a right angle at the bottom (windmill, road sign, etc.) is at the top of the right angle.

d. 几种几何图形组成的符号，如气象站、雷达站、无线电杆等，定位点在下方图形的中心点或交叉点。

d. For a symbol consisting of several geometries, such as the weather station, radar station and radio pole, the positioning point is in its center.

e. 下方没有底线的符号，如窑、亭、山洞等，定位点在其下方两端点间的中心点。非比例符号除在简要说明中规定按真方向表示外，其他的均垂直于南图廓方向描绘。

e. For a symbol without a bottom line, such as the kiln, pavilion and cave, the positioning point is in its center. Except that the non-proportional symbols as specified in the brief description are depicted in the true direction, others are depicted perpendicular to the south neatline.

③线形符号（半依比例符号）。线形符号是指长度依地形图比例尺缩绘，而宽度不依比例尺缩小表示的狭长的地物符号。用线形符号表示的地物常常是一些带状延伸地物，如铁路、公路、通信线、电线等。线形符号的中心线一般表示其实地地物的中心位置，但是城墙和篱笆等地物中心位置在其符号的底线上。

③ Line symbol (partial proportional symbol). A line symbol is a long and narrow symbol whose length is scaled down at the map scale with the width kept unchanged. Ground objects represented by linear symbols are often strip-shaped extensions, such as railways, highways, communication lines, and electric wires. The centerline of a linear symbol generally indicates the center position of a ground object, but the center position of a ground object such as the rampart and fence is on the bottom line of the symbol.

④地物注记。使用文字、数字或特定的符号对地物加以说明或补充，这种称为地物注记。地物注记分为文字注记、数字注记和符号注记三种，如居民地、山脉、河流名称，河流的流速、深度，房屋的层数、控制点高程、植被的种类、水流的方向等。

④ Notes of ground object. Notes of a ground object are characters, numbers or specific symbols that describe or supplement information on the ground object. Notes are categorized into character notes, numeral notes and symbol notes, such as names of residential areas, mountains and rivers, flow velocity and depth of rivers, number of floors of houses, elevation of control points, types of vegetation and direction of water flow.

### （2）地貌符号

### (2) Symbols of land features

地貌形态多种多样，可按其起伏变化的程度分成平地、丘陵地、山地、高山地。图上表示地貌的方法有多种，对于大、中比例尺地形图主要采用等高线。对于特殊地貌，则采用特殊符号表示。

Landforms are categorized into flat, hill, mountain and alpine land based on the degree of undulating. There are various methods to represent the land feature on the map, and contour lines are mainly used for large-scale and medium-scale topographic maps. Special land features are represented by special symbols.

等高线是地面上高程相等的相邻点连接而成的闭合曲线。如图 5-4 所示，设有一座小山头的山顶被水恰好淹没时的水面高程为 110 m，水位每退 10 m，则坡面与水面的交线即为一条闭合的等高线，其相应高程 100 m、90 m、80 m。将地面各交线垂直投影在水平面上，按一定比例尺缩小，从而得到一表现山头形状、大小、位置以及其起伏变化的等高线。

Contour lines are closed curves formed by connecting adjacent points with the same elevation on the ground. As shown in Figure 5-4, we assume that the elevation of the water surface is 110 m when the top of the hill is just submerged by water. When the water level drops by 10 m, the intersection lines between the slope surface and the water surface are closed contour lines with corresponding elevations of 100 m, 90 m and 80 m. A contour line showing the shape, size, position and undulating degree of the hill can be obtained by vertically projecting the intersection lines on the ground on the horizontal plane and scaling down at a certain scale.

相邻等高线之间的高差称为等高距或等高线间隔。在同一幅地形图上，等高距是相同的。相邻等高线间的水平距离 $d$ 称为等高线平距。$d$ 越大，表示地面坡度越缓；反之，则越陡。相邻等高线之间的地面坡度为

The difference in elevation between any

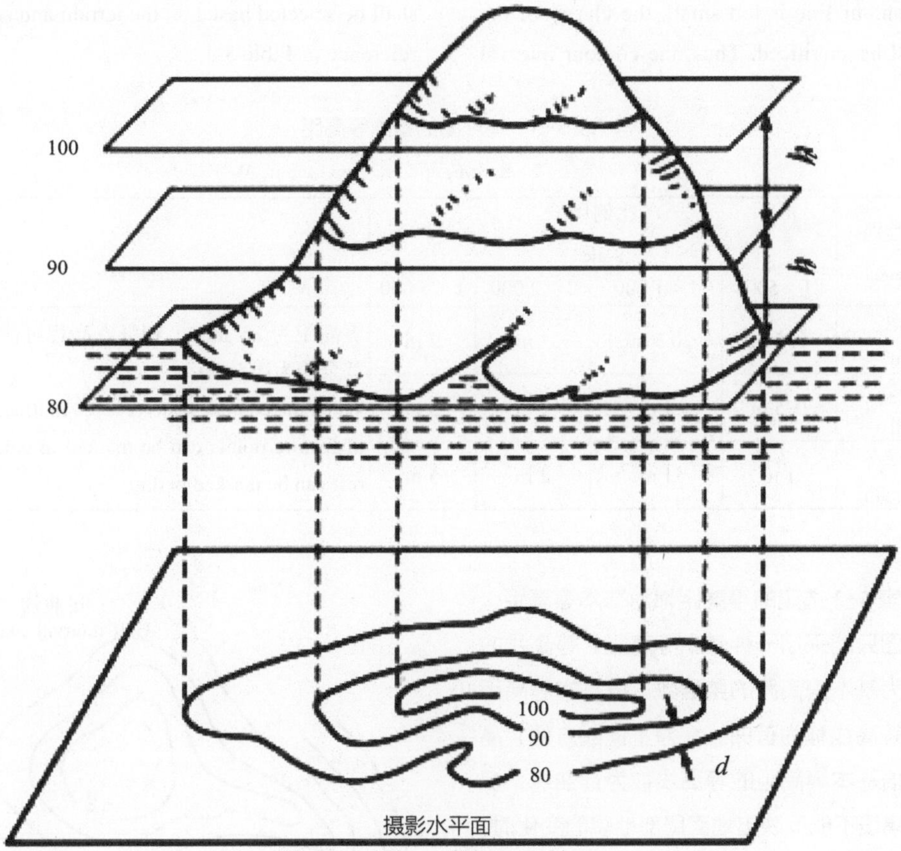


图5-4 等高线

Figure 5-4 Contour Line

two consecutive contour lines is called a contour interval. The contour interval is the same on the same topographic map. The horizontal distance $d$ between any two consecutive contour lines is called the horizontal distance. The larger d is, the gentler the ground slope is; conversely, the steeper it is. The ground gradient between two consecutive contour lines is represented as

$$i=\frac{h}{dM} \qquad (5.3)$$

式中 $M$ 是地形图的比例尺分母。

Where, $M$ is denominator of the map scale.

坡度与平距成反比。在同一幅地形图上，等高线平距越大，表示地貌的坡度越小；反之，坡度越大。因此，可根据图上等高线的疏密程度判断地面坡度的陡缓。

The gradient is inversely proportional to the horizontal distance. On the same topographic map, the larger the horizontal distance, the smaller the gradient of the land feature; on the contrary, the larger the slope. Therefore, the ground gradient can be identified based on the density of the contour lines on the map.

用等高线表示地貌，等高距选择过大，就不能精确显示地貌；反之，选择过小，等高线密集，会失去图面的清晰度。因此，应根据地形和比例尺，参照表 5-3 选用等高距。

Contour lines are used to represent the land feature. If the contour interval is too large, the land feature cannot be displayed accurately; otherwise, if the contour line is too small, the clarity of the map will be sacrificed. Thus, the contour interval shall be selected based on the terrain and scale with reference to Table 5-3.

表 5-3 地形图的基本等高距

Table 5-3 Basic Contour Interval of Topographic Map

| 地形类别<br>Terrain | 比例尺<br>Scale | | | | 备注<br>Remarks |
|---|---|---|---|---|---|
| | 1 : 500 | 1 : 1 000 | 1 : 2 000 | 1 : 5 000 | |
| 平地<br>Flat | 0. 5 m | 0.5 m | 1 m | 2 m | 等高距为0.5 m时，特征点高程可注至cm，其余均为注至dm。<br>When the contour interval is 0.5 m, the elevation of feature points can be marked in cm, and the rest can be marked in dm. |
| 丘陵<br>Hill | 0. 5 m | 1 m | 2 m | 5 m | |
| 山地<br>Mountain | 1 m | 1 m | 2 m | 5 m | |

按表 5-3 选定的等高距称为基本等高距，同一幅图只能采用一种基本等高距。等高线的高程应为基本等高距的整倍数。按基本等高距描绘的等高线称为首曲线；为了读图方便，高程为 5 倍基本等高距的等高线称为计曲线；在基本等高线不能反映出地面局部地貌的变化时，可用二分之一基本等高距的等高线，称为间曲线（图 5-5）。

The contour intervals listed in Table 5-3 are basic contour intervals, and only one basic contour interval can be used in one map. The elevation of a contour line shall be an integral multiple of the basic contour interval. A contour line at basic contour intervals is an intermediate contour; for the convenience of reading, a contour line with an elevation of 5 times the basic contour interval is called an index contour; when basic contour lines cannot show the change of local land features of the ground, contour lines with half of the basic contour interval can be used, and they are called half-interval contours（Figure 5-5）.

更加细小的变化还可用四分之一基本等高距的等高线，称为辅助等高线，亦称助曲线。

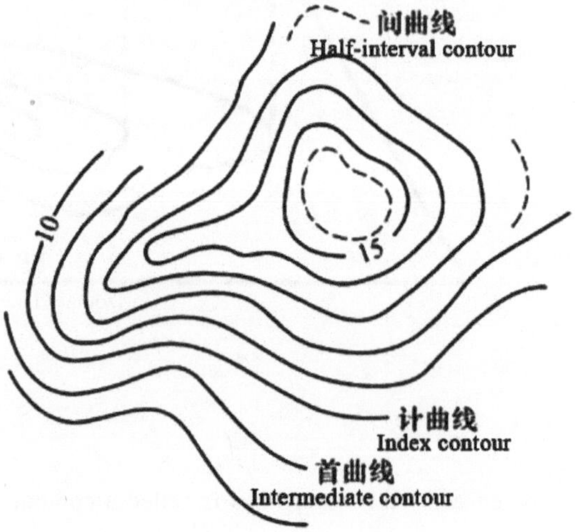


图5-5 等高线

Figure 5-5 Contour Lines

Contour lines represented with dense short-dashed lines with a quarter of the basic contour interval can be used to show more details of land features, and they are called auxiliary contours.

①等高线的特性。

① Contour features

a. 等高性：位于同一条等高线上各点的高程都相等。

a. Equal elevation: All points on a contour line are of the same elevation.

b. 闭合性：等高线是一条闭合的曲线，它若不在本幅图内闭合，就延伸或迂回到其他图幅内闭合。

b. Close: All contour lines must close either within the map boundary or outside.

c. 非交性：不同高程的等高线不能相交也不能重合，只有通过绝壁处的等高线才会重合，通过悬崖时等高线才会相交。

c. No cross: Contour lines of different elevations cannot cross each other but they can unite to form one line only in the case of a vertical cliff. If contour lines of different elevations cross each other, it shows the existence of overhanging cliffs.

d. 密陡稀缓性：等高距一定时，等高线越密，表示地势越陡；等高线越稀，表示地势越平缓。

d. Contours dense, slope steep; contours sparse, steep gentle: When the contour distance is constant, if the contour lines are spaced very closely then they represent and indicate a steep slope. If the lines are spaced far apart then it represents a gentle slope.

e. 正交性：等高线通过分水线时，与分水线垂直相交，并向低处凸出；等高线通过集水线时，与集水线垂直相交，并向高处凸出。

e. Orthogonality: Contour lines cross watershed lines vertically, with the convex point curving to the lower ground; contour lines cross watercourse lines vertically, with the convex point curving to the higher ground.

②几种典型地貌的等高线。地球表面高低起伏的形态千变万化，但它们都是由以下几种典型的地貌综合而成的，如图 5-6 所示。

② Contours of typical land features. The features of the earth's surface vary greatly but can be categorized into the following typical land features, as shown in Figure 5-6.

a. 山头和洼地的等高线。

a. Contours of hills and depressions.

山头的等高线由外圈向内圈高程逐渐增加 [ 图 5-7(a)], 洼地的等高线由外圈向内圈高程逐渐减小 [ 图 5-7(b)], 这样就可以根据高程注记区分山头和洼地，也可以用示坡线来指示斜坡向下的方向。在山头、洼地的等高线上绘出示坡线，有助于地貌的识别。

The contour lines of hilltop form circles, whereby the values of their elevation increase from the edge to the center [Figure 5-7(a)]. The contour lines of depression also form circles. The values of their elevation, however, decrease from the edge to the center[Figure 5-7(b)]. Thus, the elevation notes can be used to distinguish depressions from hilltops. Slope lines may also be added on contour lines of hilltops and depressions to indicate the downward direction of the slopes, so as to identify land features.

b. 山脊和山谷的等高线。

b. Contours of ridges and valleys.

山坡的坡度和走向发生改变时，在转折处就会出现山脊或山谷地貌。山脊的等高线均向下坡方向凸出，两侧基本对称。山脊线是山体延伸的最高棱线，也称分水线。山谷的等高线均凸向高处，两侧也基本对称。山谷线是谷底点的连线，也称集水线，如图 5-8 所示。

When the gradient and strike of a hill change, ridges or valleys will appear at the turning point. Contour lines of ridges are basically symmetrical on both sides, with the convex point curving in the downhill direction. A ridge line is the top edge of a mountain chain, also known as the watershed line. Contour lines of valleys are also basically symmetrical on both sides, with the convex point curving to the higher ground. A valley line is a line joining the lowest point of a valley, also known as the watercourse line, as shown in Figure 5-8.

c. 鞍部。

c. Saddle.

相邻两个山头之间呈马鞍形的低凹部分称为鞍部。鞍部是山区道路选线的重要位置。鞍部左、右两侧的等高线是近似对称的两组山脊线和两组山谷线，如图 5-9 所示。

A saddle is a dip or low point between two hills. Saddle is crucial for route selection in mountainous areas. Two sets of ridge lines and two sets of valley lines are approximately symmetrical on both sides of the saddle, as shown in Figure 5-9.

d. 陡崖和悬崖。

d. Escarpment and cliff.

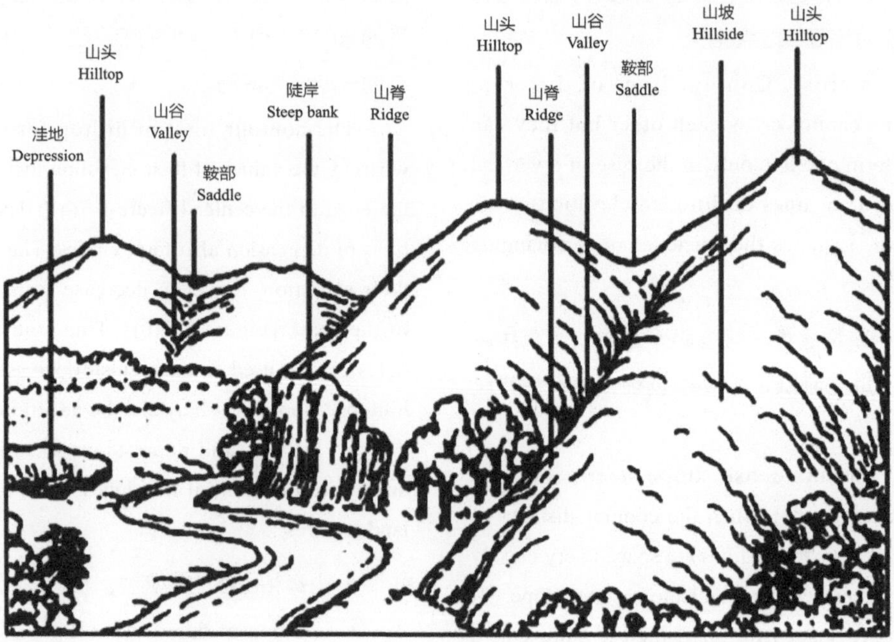


图5-6　典型的地貌

Figure 5-6 Typical Land Features

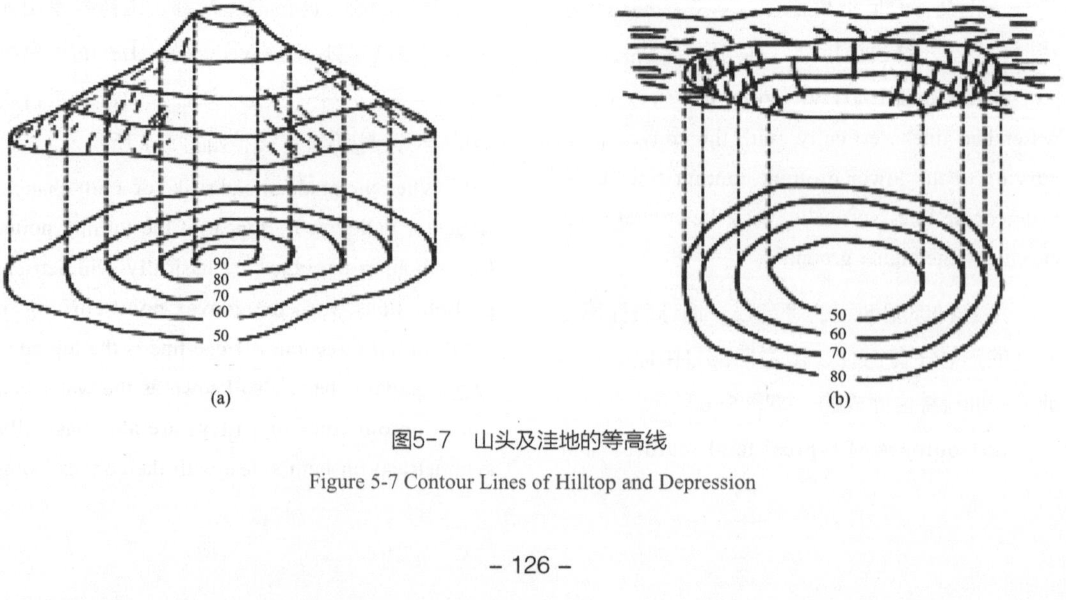


图5-7　山头及洼地的等高线

Figure 5-7 Contour Lines of Hilltop and Depression

陡崖是坡度在 70° 以上的陡峭崖壁，有石质和土质之分。如果用等高线表示，则非常密集或重合为一条线，因此采用陡崖符号来表示，如图 5-10(a)、(b) 所示。

An escarpment is a steep slope with a gradient of more than 70°, which is categorized into stone escarpment and soil escarpment. Contour lines of an escarpment are dense or united to form one line and are represented by symbols as shown in Figure 5-10 (a) and (b).

悬崖是上部凸出、下部凹进的陡崖。悬崖上部的等高线投影到水平面时，与下部的等高线相交，下部凹进的等高线部分用虚线表示，如图 5-10(c) 所示。

A cliff is the escarpment convex in the upper part and concave in the lower part. The contour lines of the upper part of the cliff that is projected to the horizontal plane intersect with the contour lines of the lower part, and the contour line of the concave lower part is represented by a dashed line, as shown in Figure 5-10 (c).

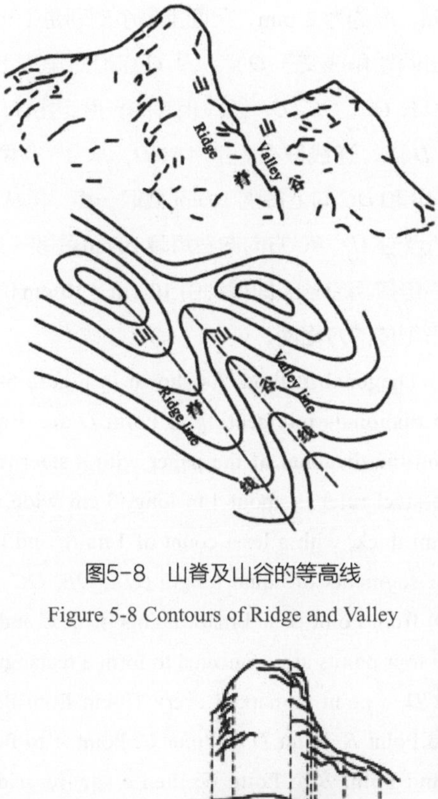


图5-8　山脊及山谷的等高线

Figure 5-8 Contours of Ridge and Valley

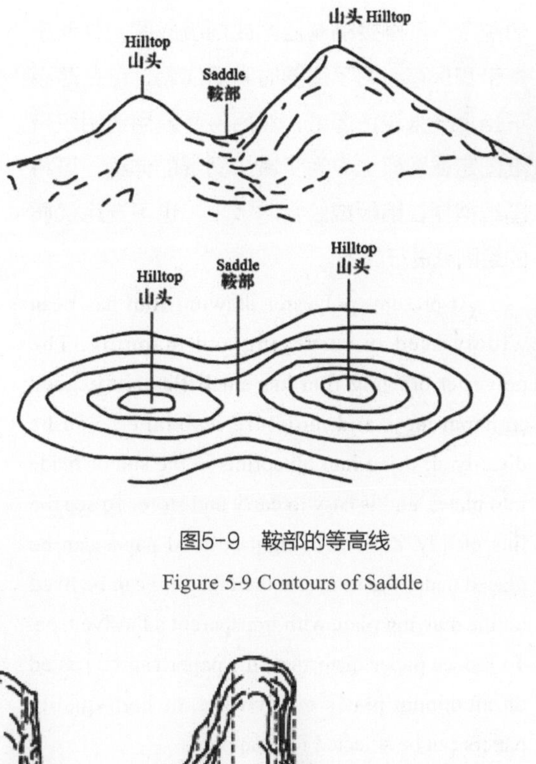


图5-9　鞍部的等高线

Figure 5-9 Contours of Saddle

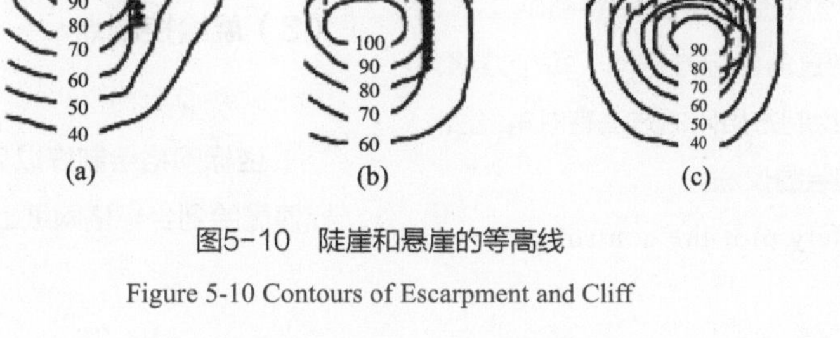


图5-10　陡崖和悬崖的等高线

Figure 5-10 Contours of Escarpment and Cliff

# 5.2 大比例尺地形图测绘

# 5.2 Surveying and Mapping of Large-scale Topographic Maps

## 5.2.1 测图前的准备工作

## 5.2.1 Preparations

### （1）图纸的准备

### (1) Preparation of map paper

目前，测绘地形图已广泛使用聚酯绘图薄膜，聚酯绘图薄膜的伸缩性小，透明度好，不怕潮湿，可直接晒制蓝图或制版成图，并便于携带和保存。为了测图时看清线画，可在薄膜下垫上一张浅色薄纸，然后将薄膜用透明胶带纸固定在图板上。为了减少图纸的伸缩，可将图纸裱糊在铝板或胶合木板上，也可选用优质的绘图纸进行测图。

At present, polyester drawing film has been widely used in surveying and mapping. The polyester drawing film has small flexibility, good transparency, and moisture resistance, can be directly air-dried into blueprints in the sun or made into plates, and is easy to carry and store. To see the line clearly, a piece of light-colored paper can be placed under the film, and then the film can be fixed on the drawing plate with transparent adhesive tape. To reduce paper distortion, the paper can be pasted on aluminum plates or plywood, or high-quality papers can be selected for mapping.

### （2）坐标格网的绘制

### (2) Drawing of coordinate grid

为了准确地将控制点展绘到图纸上，首先要精确地绘制直角坐标方格网，每个方格为 10 cm × 10 cm，绘制方格网的方法有对角线法、坐标格网尺法和绘图仪法。

To accurately plot the control points on the paper, it is necessary to accurately draw the rectangular coordinate system first, with each square grid being 10 cm×10 cm. The square grid can be drawn with the diagonal method, grid scale method and plotter method.

对角线法。如图 5-11 所示，沿图纸的对角，用一钢板尺 ( 钢板尺长约为 1 m，宽约为 3 cm，厚约为 2 mm，尺面上最小刻划是 1 mm) 绘出两对角线交于 *O* 点，从 *O* 点沿对角线上量取 *OA*、*OB*、*OC*、*OD* 四段相等的长度定出 *A*、*B*、*C*、*D* 点，连线取得矩形 *ABCD*，从 *A*、*D* 两点沿 *AB* 和 *DC* 向上每隔 10 cm 截取一点，再从 *A*、*B* 两点沿 *AD*、*BC* 方向向右每隔 10 cm 截取一点；而后连接相应各点即得到由 10 cm × 10 cm 的正方形组成的方格网。

Diagonal method. As shown in Figure 5-11, two diagonals intersecting at Point *O* are drawn along the diagonal of the paper with a steel ruler (the steel ruler is about 1 m long, 3 cm wide and 2 mm thick, with a least count of 1 mm), and take four segments of equal length (*OA*, *OB*, *OC* and *OD*) from Point *O* and mark points *A*, *B*, *C* and *D*. The four points are connected to form a rectangular *ABCD*, a point is marked every 10 cm from Point *A* to Point *B*, Point *D* to Point *C*, Point *A* to Point *D* and Point *B* to Point *C*; then a square grid of 10 cm×10 cm can be formed by connecting the corresponding points.

### （3）展绘控制点

### (3) Plotting of control points

坐标网格绘制好以后，可将控制点按其坐标值展绘到坐标格网纸上，这项工作称为展点。

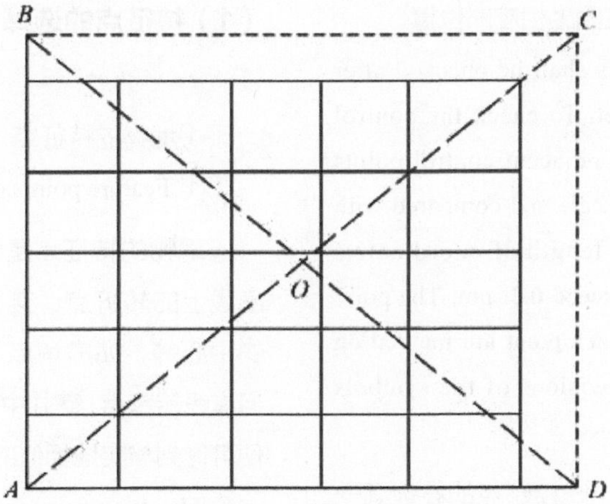


图5-11　对角线法

Figure 5-11 Diagonal Method

展绘控制点前，首先按图的分幅位置将坐标格网线的坐标值在相应方格网边线的外侧（图 5-12）标注。展点时，先根据控制点的坐标确定其所在的方格。图 5-12 中控制点 $A$ 的坐标为 $x_A$= 634.30 m，$y_A$= 636.15 m，由其坐标可知 $A$ 点的位置在 *lpnm* 方格内。再按 $y$ 坐标值分别从 $l$、$p$ 点按测图比例尺向右各量 36.15 m，得 $a$、$b$ 两点。同法，从 $p$、$n$ 点向上各量 34.30 m，得 $c$、$d$ 两点，连接 $ab$ 和 $cd$，其交点即为 $A$ 点的位置。用同样的方法将图幅内其余控制点展绘在图纸上，各点的符号应绘出。

After the coordinate grid is completed, control points can be plotted on the coordinate grid paper based on the coordinates, and the process is called point plotting. Before plotting the control points, mark the coordinates of the grid on the outer side of the square grid line based on the sheet subdivision (Figure 5-12). The position of a point is based on the coordinates of the control point. The coordinates of control point $A$ in Figure 5-12 are $x_A$= 634.30 m and $y_A$= 636.15 m, then the position of Point $A$ is within square *lpnm*. Points $a$ and $b$ can be obtained by measuring 36.15 m from point $l$ and point $p$ to the right at the mapping scale. In the same way, points $c$ and $d$ can be obtained by measuring 34.30 m upwards from points $p$ and $n$. By connecting Point $a$ with Point $b$ and Point $c$ with Point $d$, we can obtain the intersection point $A$. Then the remaining control points are plotted on the sheet and indicated with symbols.

控制点展绘好后，应进行校核。校核方法是：用比例尺量出各相邻控制点之间的长度，与坐标反算长度相比较，其差值不应超过图上 0.3 mm。检查无误后，按地形图图式的规定将

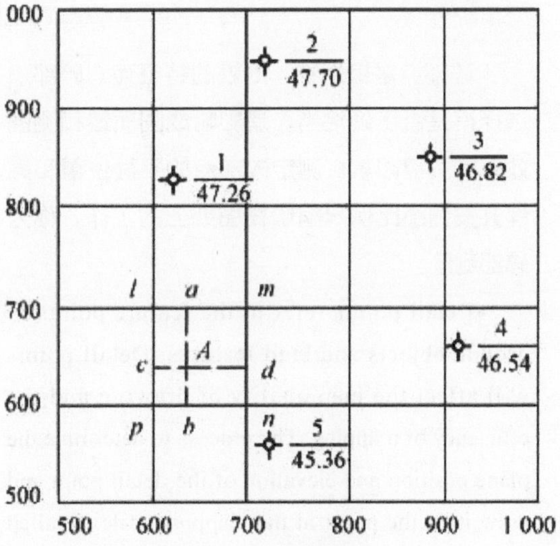


1 : 1 000

图5-12　控制点展绘

Figure 5-12 Plotting of Control Points

各点的点号和高程标注在图上相应的位置。

All the control points shall be checked after being plotted on the sheet. To check the control point, the length between adjacent control points shall be measured with a scale and compared with the inversely-calculated length of coordinates. The difference shall not exceed 0.3 mm. The point number and elevation of each point are marked on the map based on the provisions of the symbols after checking for correctness.

展点经检查合格后，在控制点的右边画一条 5 mm 的横线，上方注明点号（或点名），横线下标注高程，并将控制点按规定的符号整饰或绘“＋”标字形符号表示。

After checking the point plotting for correctness, a 5 mm horizontal line is drawn on the right side of the control point, with the point number (or point name) indicated above and the elevation marked below the horizontal line. A specified symbol or “+” symbol is added or drawn to represent the control point.

## 5.2.2 量角器配合经纬仪测图法

## 5.2.2 Mapping with Protractor and Theodolite

碎部点是指地物、地貌的特征点。碎部点选择得是否正确恰当，是影响成图质量和测图效率的关键因素。测定碎部点的平面位置和高程并按测图比例尺缩绘在图纸上的工作，称为碎部测量。

Detail points refer to the feature points of ground objects and land features. Detail points will affect the map quality of drawing and the efficiency of mapping. The process to determine the plane position and elevation of the detail point and draw it on the paper at the mapping scale is called surveying of detail points.

### （1）特征点的选择

### (1) Selection of feature points

①地物的特征点。

① Feature points of ground objects.

地物的特征点是指决定地物形状的地物轮廓线上的转折点、交叉点、弯曲点及独立地物的中心等，如房角点、道路转折点、交叉点、河岸线转弯点、窨井中心点等。连接这些特征点，便可得到与实地相似的地物形状。

The feature points of ground objects refer to the turning points, intersection points, bending points and centers of detached ground objects that determine the shape of ground objects, such as corner points of houses, turning points of roads, intersection points, turning points of river bank lines, and center points of drainage pits. By connecting these feature points, we can get the shape of ground objects similar to the real ones.

②地貌的特征点。

② Feature points of land features.

地面上的各种地形虽然十分复杂，但可以看成是由向着各个方向倾斜和具有不同坡度的面组成的多面体。而山脊线、山谷线、山脚线等地性线是多面体的棱线。因此，地貌的特征点应选在这些地形线的转折点（方向变化和坡度变化处）上。另外，还应选择山头、鞍部、洼坑底部等处，如图 5-13 所示。根据这些特征点的高程勾绘等高线，即可将地貌在图上表示出来。在地面平坦的地方或坡度无明显变化的地区，碎部点的间距和碎部点的最大视距应符合表 5-4 的规定。城市建筑区的最大视距见表 5-5。

Although there are different land features on the ground, they can be regarded as polyhedra composed of surfaces tilted in different directions and with different gradients. The ridge line, valley line, foot line and other terrain lines can be regarded as edge lines of polyhedra. Therefore, the feature points of the land feature shall be set at the turning points (change in direction and gradient) of terrain lines. Feature points shall also be located at the hilltop, saddle, pit bottom, etc., as shown in Figure 5-13. A land feature can be shown on a map by drawing contour lines based on the elevation of feature points. In areas where the ground is flat or where there is no significant change in gradient, the spacing between detail points and the maximum sight distance for detail points shall be as specified in Table 5-4. The maximum sight distance in urban building area is shown in Table 5-5.

图5-13　地貌图

Figure 5-13 Land Features

表 5-4　平坦地区碎部点最大视距

Table 5-4 Maximum Sight Distance for Detail Points in Flat Area

| 测图比例尺 Mapping scale | 地形点最大间距/m Maximum spacing between topographical points/m | 最大视距/m Maximum sight distance/m | |
|---|---|---|---|
| | | 主要地物点 Main planimetric points | 次要地物点和地形点 Secondary planimetric points and topographical points |
| 1：500 | 15 | 60 | 100 |
| 1：1 000 | 30 | 100 | 150 |
| 1：2 000 | 50 | 180 | 250 |
| 1：5 000 | 100 | 300 | 350 |

表 5-5　城市建筑碎部点最大视距

Table 5-5 Maximum Sight Distance for Detail Points of Urban Buildings

| 测量比例尺 Measuring scale | 最大视距/m Maximum sight distance/m | |
|---|---|---|
| | 主要地物点 Main planimetric points | 次要地物点和地形点 Secondary planimetric points and topographical points |
| 1 ∶ 500 | 15（量距） 15 (measured distance) | 70 |
| 1 ∶ 1 000 | 80 | 120 |
| 1 ∶ 2 000 | 120 | 200 |

## （2）碎部测量

### (2) Surveying of detail points

经纬仪测绘法是将经纬仪安置在测站上，测定碎部点的方向与已知方向之间的夹角，并用视距测量方法测出测站点至碎部点的平距及碎部点的高程。绘图板安置于测站旁，根据测定数据，用量角器（又称半圆仪）和比例尺把碎部点的平面位置展绘在图纸上，并在点的右侧注明其高程。最后再对照实地描绘地物和地貌。一个测站上的测绘工作步骤如下。

The method of surveying and mapping with theodolite is to place the theodolite on the observation station, measure the angle between the direction of the detail point and the known direction, and measure the horizontal distance from the observation station to the detail point and the elevation of the detail point by the tacheometric surveying method. The drawing board is placed beside the observation station and the plane position of the detail point is plotted on the drawing with a protractor (also known as a graphometer) and a scale based on the measured data, with the elevation indicated on the right side of the point. Finally, the ground objects and land features are depicted based on the real ones. The surveying and mapping steps on an observation station are as follows.

①安置仪器。如图 5-14 所示，将经纬仪安置于测站点 $A$ 上，对中、整平，并量出仪器高度 $i$。

① Place the instrument: As shown in Figure 5-14, place the theodolite on observation point $A$, center and level it, and measure the elevation $i$ of the instrument.

②定向。用经纬仪盘左位置瞄准另一控制点 $B$，设置水平度盘读数为 0° 0′ 0″。$B$ 点称为后视点，$AB$ 方向称为起始方向或后视方向。在小平板上固定好图纸，并安置在测站附近，注意使图纸上控制边方向与地面上相应控制边方向大致相同。连接图上对应的控制点 $a$、$b$，并适当延长 $ab$ 线，$ab$ 为图上起始方向线。然后用小针通过量角器圆心插在 $a$ 点，使量角器圆心 $a$ 固定在点上。

② Determine the direction: Aim at another control point $B$ with the circle of the theodolite placed on the left side of the telescope, and set

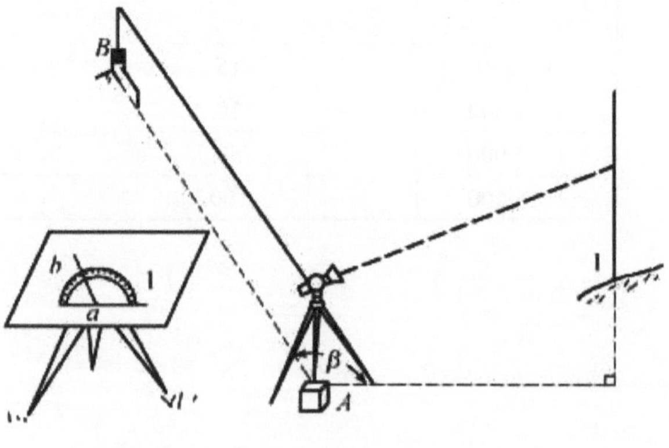


图5-14　碎部测量

Figure 5-14 Surveying of Detail Points

the reading of the horizontal circle to 0°00′00″. Point *B* is called the backsight point, and the *AB* direction is called the starting direction or the backsight direction. Fix the paper on the board and place it near the observation station. Note that the direction of the control edge on the map is roughly the same as that on the ground. Connect the corres-ponding control points *A* and *B* on the map, and appropriately extend the *ab* line, which is the starting direction line on the map. Then insert a small needle through the center of the protractor at point a to fix the center of the protractor at point *a*.

③立尺。立尺员将视距尺依次立在地物和地貌的特征点上。立尺前，立尺员应根据实地情况及本测站实测范围，按照“概括全貌、点少、能检核”的原则选定立尺点，并与观测员、绘图员共同商定跑尺路线。比如在平坦地区跑尺，可由近及远，再由远及近地跑尺，立尺结束时处于测站附近。在丘陵或山区，可沿地形线或等高线跑尺。

③ Set the stadia rod: The surveyor set the stadia rod on the feature points of the ground object and the land feature in sequence. Before setting the stadia rod, the surveyor shall select the setting points according to the field situation, the surveying range of the observation station, and the principle of “reflecting the whole picture with fewer points that are easy to check”, and jointly agree on the surveying route with the observer and the drafter. For example, surveying can be carried out from near to far and then from far to near in a flat area, with the stadia rod near the observation station at the end. In hilly or mountainous areas, surveying can be carried out along the terrain line or contour line.

④观测。观测员转动经纬仪照准部，瞄准1点视距尺，读尺间隔(上丝读数、下丝读数)、中丝读数 *n*、竖盘读数及水平角。为了计算方便，往往下丝对准一个整刻划，这样一次即可读出上、下丝之差的尺间隔。同法观测周围2、3等各点。

④ Observation: The observer rotates the alidade part of the theodolite to aim the stadia rod at Point 1 and reads the scale interval (upper crosshair reading and lower crosshair reading), middle crosshair reading *n*, vertical circle reading and horizontal angle. For the convenience of calculation, the lower crosshair is often aligned with a whole scale mark, so that the difference (scale interval) between the upper crosshair reading and the lower crosshair reading can be read at one time. Then surrounding points such as points 2 and 3 are observed with the same method.

⑤记录与计算。记录员将测得的尺间隔、中丝读数、竖盘读数及水平角等数据依次填入地形碎部测量手簿（表5-6）中。然后根据测得数据按视距测量计算公式计算水平距离 *D* 和高程 *H*。

⑤ Records and calculations: The recorder fills the measured data such as scale interval, middle crosshair reading, vertical circle reading and horizontal angle in the topographic survey recorder (Table 5-6) in sequence and then calculates the horizontal distance *D* and elevation *H* based on the measured data and the calculation formula for sight distance measurement.

表 5-6　地形碎部测量手簿

Table 5-6 Details Survey Recorder of Terrain

| 测站：$A$　　后视点：$B$<br>Observation Station: $A$　　Rear View Point: $B$<br>测站高程 $H_A$ = 27.40 m<br>Station elevation $H_A$ = 27.40 m | | | | $i$=1.42 m | | | | |
|---|---|---|---|---|---|---|---|---|
| 点号<br>Point number | 视距 $KL$/m<br>Sight Distance $KL$/m | 中丝读数 $n$/m<br>Middle crosshair reading $n$/m | 竖盘读数<br>Vertical circle reading | 竖直角<br>Vertical angle | 水平角<br>Horizontal angle | 水平距离 $D$/m<br>Horizontal distance $D$/m | 高程 $H$/m<br>Elevation $H$/m | 备注<br>Remarks |
| 1 | 76.0 | 1.42 | 93° 28′ | −3° 28′ | 114° 00′ | 75. 7 | 22. 81 | |
| 2 | 51.4 | 1. 55 | 91° 45′ | −1° 45′ | 172° 40′ | 51. 4 | 25. 70 | 房角<br>House corner |
| 3 | 37.5 | 1. 60 | 93° 00′ | −3° 00′ | 327° 36′ | 37. 4 | 25. 26 | 电杆<br>Electric pole |
| 4 | 25.7 | 2. 42 | 87° 26′ | +2° 34′ | 16° 24′ | 25. 7 | 27. 5 | |

⑥刺点。绘图员转动量角器，将量角器上等于水平角值（如碎部点 1 的水平角 =114°）的刻划线对准起始方向线，如图 5-15 所示。此时量角器的零方向便是碎部点 1 的位置，用铅笔在图上标定，并在点的右侧注明其高程。用同样的方法将其余各碎部点的平面位置及高程绘于图上。仪器搬到下一站时，应先观测前站所测的某个明显碎部点，以检查由两站测得该点的平面位置和高程是否相符。如相差较大，应查明原因，纠正错误，再继续进行测绘。

⑥ Prick point: The draftsman rotates the protractor and aligns the scale mark on the protractor equal to the horizontal angle value (e.g. horizontal angle of Detail Point 1 = 114°) with the starting direction line, as shown in Figure 5-15. At this time, the zero direction of the protractor is the position of Detail Point 1, which is marked with a pencil on the map and its elevation is indicated on the right side of the point. The plane position and elevation of other detail points are indicated on the map in the same way. When the instrument is moved to the next station, an obvious detail point measured at the previous station shall be observed first to check whether the plane position and elevation of the point measured at the two stations are consistent. In case of a significant difference, the surveying and mapping shall be continued only after the cause is found out and the error is corrected.

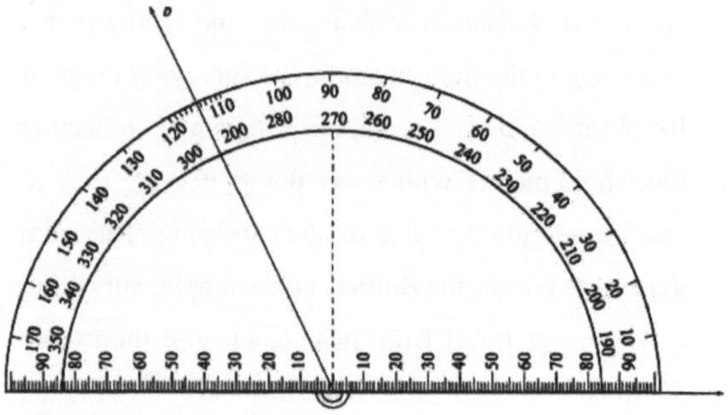

图5-15　刺点

Figure 5-15 Prick Point

### （3）碎部测量的注意事项

(3) Precautions for detail points surveying

①测图过程中，每观测 20~30 个碎部点后，应检查起始方向，其归零差不得超过 4′。否则，应重新定向，并检查已测的碎部点。

① During mapping, the starting direction shall be checked after every 20-30 detail points are observed, and the misclosure of round shall not exceed 4′. Otherwise, the direction shall be determined again and the measured detail points shall be checked.

②立尺人员应将视距尺竖直，并随时观察立尺点周围的情况，弄清楚碎部点之间的关系；地形复杂时还需绘出草图，以协助绘图人员做好绘图工作。绘图人员要注意图面信息正确、整洁、注记清晰，并做到碎部点测量、展绘后立即检查。

② The surveyor shall set the stadia rod upright, and observe the situation around the setting point at any time to grasp the relationship between the detail points. In case of complex terrain, a sketch shall be drawn to assist the draftsman in the drawing. The draftsman shall make sure that the information on the sheet is correct, well organized and clearly marked. The check should be carried out immediately after the surveying and the plotting of detail points.

③当该站工作结束时，应检查有无漏测、错测，并将图面上的地物、地形线及等高线与实地对照，及时发现问题，予以纠正。

③ When the work at a station is completed, it is necessary to check whether there is any missed point or error and compare the ground objects, terrain lines and contour lines on the map with the field to find out the problems in time and correct them (if any).

④若采用光电测距仪测绘地形图，其测绘方法与经纬仪测绘法基本相同，所不同的是用光电测距来代替视距法测距，故用单棱镜取代塔尺。

④ If the topographic map is surveyed with a geodimeter, the surveying and mapping method is basically the same as that of the theodolite, but the electro-optical distance measurement method is used instead of the tacheometric method, so the telemeter rod is replaced by a single prism.

## 5.2.3 平板仪测图法

## 5.2.3 Mapping with Plane Table

### （1）平板仪的构造

### (1) Construction of plane table

平板仪由平板、照准仪和若干附件组成，如图 5-16 所示。平板部分由图板、基座和三脚架组成。基座用中心螺旋安装在三脚架上，拧松中心螺旋，平板可在脚架头上作小范围移动。基座下部有脚螺旋可以整平图板。另外，还装有制动和微动螺旋，用于控制图板在水平方向的转动。

A plane table consists of a table, an alidade and accessories, as shown in Figure 5-16. The table consists of a drawing board, a base and a tripod. The base is installed on the tripod with connecting screw on the center and the table can move in a small range on the tripod head by loosening the screw. The lower part of the base has foot screws to level the drawing board. There are also stop screws and tangent screws to control the rotation of the

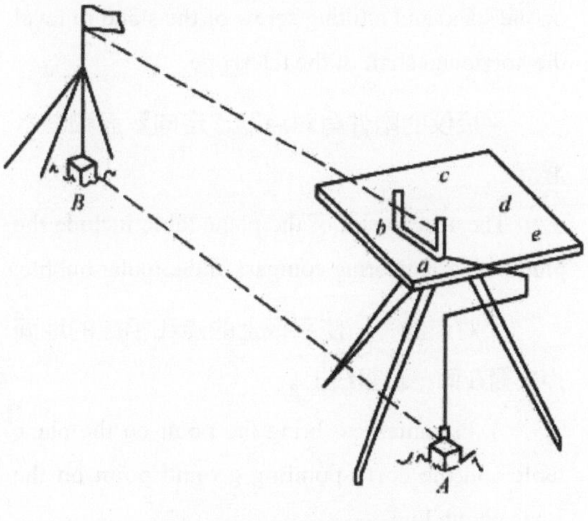


图5-16　小平板及附件

Figure 5-16 Small Plane Table and Accessories

drawing board in the horizontal direction.

照准仪主要是由望远镜、竖盘和直尺组成。望远镜和竖盘相当于经纬仪的视距测量部分，配合在照准点上竖立的视距标尺，可以测定距离和高差；直尺和望远镜的视准轴在同一竖直面内，望远镜瞄准目标后，根据直尺在平板上划出的方向线即代表瞄准方向；望远镜有垂直方向的制动螺旋、微动螺旋、物镜对光螺旋、目镜对光螺旋和竖盘水准管微动螺旋；望远镜的支柱上还有横向水准管及支柱的微倾螺旋，用以置平望远镜的横轴。

The alidade is mainly composed of a telescope, a vertical circle and a ruler. The telescope and vertical circle are equivalent to the sight distance measurement part of the theodolite, and the distance and elevation difference can be measured with the sight distance scale on the sighting point. The sight axis of the ruler and the telescope is in the same vertical plane. After the telescope aims at the target, the direction line drawn on the plane table with the ruler represents the sighting direction. In the vertical direction of the telescope, there are stop screws, tangent screws, objective lens focusing screws, eyepiece lens focusing screws and vertical tangent screws for the level tube of the vertical circle. The telescope also has a horizontal level tube on the stand and a tilting screw of the stand to level the horizontal shaft of the telescope.

平板仪的附件有对点器、定向罗盘和圆水准器。

The accessories of the plane table include the plummet, orienteering compass and circular bubble.

①对点器——使平板上的点和相应的地面点安置在同一铅垂线上。

① Plummet—to bring the point on the plane table and the corresponding ground point on the same plumb line.

②定向罗盘——用于平板仪的近似定向。

② Orienteering compass—for approximate orientation of the plane table.

③圆水准器——用以整平图板。

③ Circular bubble—to level the drawing board.

平板测图的仪器除大平板仪外，还有小平板仪。小平板仪是大平板仪简化并使之轻便的平板仪，主要部件有照准器、图板、三脚架和对点器。小平板仪的操作与使用和大平板仪相同。

Plane tables for mapping can be categorized into large and small ones by size. A small plane table is simplified and lightweight and mainly composed of an alidade, a drawing board, a tripod and a plummet. The operation method of a small plane table is the same as that of a large one.

### （2）平板仪的安置

### (2) Setting of plane table

平板仪测量的原理是相似图形的图解水平投影。所以在一个测站上，平板仪的安置包括对点、整平和定向三项工作。

The principle of plane table surveying is the graphical horizontal projection of similar patterns. Therefore, there are three steps to setting a plane table at an observation station: point alignment, leveling and orientation.

①对点。

① Point alignment.

对点就是使图板上的 $a$ 点和地面上相应的 $A$ 点位于同一铅垂线上，如图 5-16 所示。其方法是先将对点器的尖端对准图板上的 $a$ 点，然后移动脚架使垂球尖对准地面点 $A$。对点的容许误差与比例尺大小有关，一般规定为 0.05 mm × $M$，$M$ 为比例尺分母。

Point alignment is to bring point $a$ on the board and the corresponding point $A$ on the ground on the same plumb line, as shown in Figure 5-16. We need to align the tip of the plummet with point *a* on the board, and then move the tripod to align the tip of the plumb bob with point *A* on the ground. The allowable error of the point alignment is related to the scale, which is generally specified as 0.05 mm×$M$, and $M$ is the denominator of the scale.

②整平。

② Leveling.

对点以后，利用圆水准器或照准仪上的水准管使图板位于水平位置，整平方法同经纬仪整平。

After point alignment, the level tube on the circular bubble or alidade shall be used to level the drawing board, and the leveling method is the same as that of the theodolite.

③定向。

③ Orientation.

经对点、整平以后，就可以进行图板定向。如图 5-16 所示，将照准仪（器）的直尺边缘紧靠在图上已知直线上，转动图板，使照准仪（器）瞄准地面目标，然后旋紧水平制动螺旋，固定图板，此时图上控制点之间的方向与地面上控制点之间的方向完全一致，这样就完成了图板的定向工作。图板定向的正确与否对测图的精度影响很大，因此，必须细心地操作。为了防止定向发生错误，必须用另一控制点方向进行定向检查。

After point alignment and leveling, orientation of drawing board can be carried out. As shown in Figure 5-16, set the edge of the ruler of the alidade against the known straight line on the figure, rotate the drawing board, aim at the ground target with the alidade, and then tighten the horizontal stop screw to fix the drawing board. At this time, if the direction between the control points on the figure is completely consistent with that between the control points on the ground, the orientation work is finished. The correct orientation of the map plate has a great impact on the accuracy of mapping, so it must be carefully carried out. To prevent wrong orientation, it must be checked with the direction of another control point.

对点、整平和定向会相互影响，故安置平板仪一般应先大致定向、整平，然后再精确对点、整平和定向。

Since point alignment, leveling and orientation will affect each other, a rough orientation and leveling shall be carried out for the plane table before accurate point alignment, leveling and orientation.

## （3）平板仪测图

## (3) Mapping with plane table

平板仪测量是一些工程单位用于测绘大比例尺地形图的一种常用的方法。平板仪是在野外直接测绘地形图的一种仪器，它可以同时测定地面点的平面位置和高程。在平板仪测量中，水平角用图解法测定，水平距离用皮尺测量或视距测量，因此平板仪测量又称图解测量。

Plane table surveying is a common method used by some engineering organizations to map large-scale topographic maps. A plane table is used for direct surveying and mapping of topographic maps in the field and can survey the plane position and elevation of ground points at the same time. For plane table surveying, the horizontal angle is measured by graphical method, and the horizontal distance is measured by tape or sight distance. So, plane table surveying is also called graphical surveying.

平板仪测绘地形图与上述的经纬仪测绘法基本相同，所不同的是水平角用图解法测定。

The method of surveying and mapping topographic maps with a plane table is basically the same as that with a theodolite, except that the horizontal angle is determined by the graphical method.

# 5.3 数字化测图

## 5.3 Digital Surveying and Mapping

数字化测图 (digital surveying and mapping, DSM) 是以电子计算机为核心，以测绘仪器和打印机等输入、输出设备为硬件，在测绘软件的支持下，对地形空间数据进行采集、传输、编辑处理、入库管理和成图输出的一整套过程。

Digital surveying and mapping ( DSM) is a complete process of collecting, transmitting, editing, databasing, and mapping topographic spatial data by using electronic computers, input and output equipment such as mapping instruments and printers, and mapping software.

### 5.3.1 数字化测图的特点

### 5.3.1 Characteristics of Digital Surveying and Mapping

数字化测图与图解法测图相比，具有明显的优点，主要体现在以下几个方面。

Compared with graphical surveying and mapping, digital mapping has obvious advantages in the following aspects.

#### （1）测图精度高

#### (1) High accuracy

在数字化测图中，野外采集数据采用高精度电子仪器，数据的流动、展点和绘图都在机器内运行。因此，绘图精度不受比例尺的影响，减少人为错误发生的机会，提高测图精度。

For digital mapping, high accuracy electronic instruments are used to collect data in the field, and the data flow, point plotting and mapping are all completed in a machine. Therefore, the mapping accuracy is not affected by the scale, which reduces human errors and improves the accuracy of surveying and mapping.

#### （2）可实现自动化测图

#### (2) Automatic operation

数字化测图能自动记录、自动解算、自动绘图，向用图者随时提供可处理的精确、规范的数字化地形图。在自动绘图时，必须对地面点赋予三类信息，即点的三维坐标、点的属性、点间关联性。因此，数据采集和点性编码是数字测图的基础。

Digital mapping can realize automatic recording, automatic calculation and automatic drawing, and provide users with accurate and standardized digital topographic maps that can be processed at any time. During automatic mapping, the three-dimensional coordinate, attribute and correlation between points will be set to ground points. Therefore, data acquisition and coding of point attributes are the basis of digital surveying and mapping.

#### （3）易于更新测图成果

#### (3) Easy update of results

城镇的发展加速了城镇地物和地貌的变化。数字化测图的结果是以点位信息和地物属性信息存入计算机的，显著改善了纸介质地形图频繁更新困难的状况。当实地情况发生改变时，只需要输入变化部分的点位编码和坐标信息，经过编辑，即可得到更新后的新图，始终保持图面整体的实时性和完整性。

Urban development has accelerated the change of ground objects and land features. It is always difficult to update paper topographic maps frequently, while updates can be easily realized by digital surveying and mapping since the results of digital surveying and mapping are stored in the computer with point information and attribute information of ground objects. When the field situation changes, the map can be updated only by inputting and editing the point codes and coordinates of the changed parts, and the map can always maintain its integrity in real time.

### (4) 结果利用和管理便利

### (4) Convenient utilization and management of results

数字化地形图方便实行分图层管理，地形信息可无限存放，图形数据容量可不受图幅负载量的限制，从而提高测量数据的利用率，拓宽了测绘工作的服务面。

The digital topographic map is convenient for management by layers. The topographic information can be stored infinitely, and the graphic data capacity is not limited by the sheet load, thus improving the utilization rate of surveying data and broadening the service range of surveying and mapping.

### (5) 有利于地理信息系统（GIS）的建立

### (5) Helpful for establishment of GIS

地理信息系统具有方便的信息查询检索功能、空间分析功能和辅助决策功能，在国民经济各领域及人们日常生活中都有广泛的应用。GIS 的主要任务就是数据采集，而数字化测图能够提供实时性较强的基础地理信息。

The geographical information system is convenient for information inquiry and retrieval, and spatial analysis and can assist in decision-making and is widely used in national economic fields and people's daily life. GIS is mainly used for data acquisition, and digital mapping can provide basic geographical information with strong real-time performance.

## 5.3.2 数字化测图的基本作业过程

## 5.3.2 Basic Operation Process of Digital Surveying and Mapping

### (1) 数据采集

### (1) Data Collection

各种数字测图系统必须首先获取地形要素的各种信息，然后才能据此成图。地形信息包括所有与成图有关的资料，如测量控制点资料、各种地物和地貌的位置数据和属性以及有关的注记等。这些地形要素的信息通常以地形数据的形式来表达。常见的地形要素数据采集方式有以下三种。

Digital mapping systems need to collect information on terrain elements for mapping accordingly. Topographic information refers to all data related to mapping, such as information on survey control points, location and attributes of ground objects and land features, and relevant notes. The information on these terrain elements is usually expressed as terrain data. There are three common methods to collect data of terrain elements.

①用全站仪采集数据。

① Data collection with total station.

在野外用全站仪采集数据时，数字测记法有“草图法”和“编码法”两种。当采用“草图法”时，全站仪的数据采集步骤如下。

When data are collected with a total station in the field, there are two digital surveying and recording methods: “sketch method” and “coding method”. When the “sketch method” is applied, the data collection steps of the total station are as follows.

全站仪数字化测图，主要分为准备工作、数据获取、数据输入、数据处理、数据输出五个阶段。在准备工作阶段，包括资料准备、控制测量、测图准备等，与传统地形测图一样。现以实际生产中普遍采用的全站仪加电子手簿测图模式为例，介绍全站仪数字化测图从数据采集到数据编码的基本过程。

The digital mapping of the total station mainly has five stages: preparation, data collection, data input, data processing and data output. The preparation stage, including data preparation, control survey, and mapping preparation, is the same as that of traditional topographic mapping. Taking the example of mapping with a total station and data recorder commonly used in practical production, the basic process output of digital mapping with a total station, which from data collection to data coding.

a. 野外碎部点采集。

a. Data collection of detail points in the field.

一般用“解算法”进行碎部点测量采集，用电子手簿记录三维坐标 ($x$、$y$、$H$) 及其绘图信息。既要记录测站参数、距离、水平角和竖直角的碎部点位置信息，还要记录编码、点号、连接点和连接线型四种信息，在采集碎部点时要及时绘制观测草图。

Generally, the “solution method” is used for the survey and data collection of detail points, and the data recorder is used to record the three-dimensional coordinates ($x$, $y$, $H$) and their mapping information. It is necessary to record not only the station parameters, distance, and locations of detail points from the horizontal angle and vertical angle, but also the code, point number, connection point and connection line type. An observation sketch must be drawn in time when collecting data on detail points.

b. 数据传输。

b. Data transmission.

用数据通信线连接电子手簿和计算机，把野外观测数据传输到计算机中，每次观测的数据要及时传输，避免数据丢失。

The data communication line is used to connect the data recorder and the computer, and the field observation data is transmitted to the computer in time to avoid data loss.

c. 数据处理。

c. Data processing.

数据处理包括数据转换和数据计算。数据处理是对野外采集的数据进行预处理，检查可能出现的各种错误；把野外采集到的数据进行编码，使测量数据转化成绘图系统所需的编码格式。数据计算是针对地貌关系的，当测量数据输入计算机后，生成平面图形、建立图形文件、绘制等高线。

Data processing includes data conversion and calculation. Data processing is to pre-process the data collected in the field and check for possible errors. The data collected in the field are coded to convert the surveying data into the encoding format required by the mapping system. The data calculation is mainly for the relationship of land features. The survey data is input into the computer to generate the plane figure, establish the graphic file, and draw the contour lines.

d. 图形处理与成图输出。

d. Graphics processing and map output.

编辑、整理经数据处理后所生成的图形数据文件，对照外业草图，修改整饰新生成的地形图，补测、重测存在漏测或测错的地方。然后加注高程、注记等，进行图幅整饰，最后输出成图。

This step is to edit and sort out the graphic data files generated after data processing, modify the newly generated topographic map based on the field sketch, supplement and resurvey the missing or wrong parts, add elevations, notes, etc., finish the map, and output the map.

e. 数据编码。

e. Data coding.

野外数据采集，仅测定碎部点的位置并不能满足计算机自动成图的需要，必须将所测地物点的连接关系和地物类别（或地物属性）等绘图信息记录下来，并按一定的编码格式记录数据。

For field data collection, the computer cannot output the map automatically only with the position of detail points. The connection relationship between the measured points of ground objects and the mapping information such as the category (or attribute) of ground objects must be recorded in a specified coding format.

目前开发的测图软件一般是根据自身特点的需要、作业习惯、仪器设备和数据处理方法制定自己的编码规则。利用全站仪进行野外测设时，编码一般由地物代码和连接关系的简单符号组成。如代码 F0、Fl、F2 分别表示特种房、普通房、简单房（F 字为“房”的第一个拼音字母，以下类同），Hl、H2 表示第一条河流、第二条河流的点位。

The coding rules of current mapping software are generally based on the characteristics, needs, operation habits, instruments and equipment and data processing methods of the software. When the total station is used for field setting-out, the code generally consists of the code of the ground object and the simple symbol of the connection relationship. For example, codes F0, Fl and F2 represent special houses, ordinary houses and simple houses [F is the first pinyin letter of the Chinese character “房 (house)”, similar in the following context] respectively, and Hl and H2 represent the position of the first river and the second river.

②用网络 RTK（实时动态测量）技术采集数据。

② Data collection with network RTK (Real Time Kinematic) technology.

目前，GPS 网络 RTK 技术弥补了 GPS 实时差分定位 RTK 技术单独利用的缺点。其数据处理的方法有虚拟参考站法（简称 VRS）、偏导数法、线性内插法和条件平差法。其中虚拟参考站法技术最为成熟。VRS 系统集 GPS、互联网、无线通信和计算机网络管理于一身，整个系统由若干个（三个以上）GPS 固定基准站、一个 GPS 网络控制中心和多个终端用户构成。

At present, GPS network RTK technology makes up for the shortcomings of using GPS real-time differential positioning RTK technology alone. The methods of data processing include the virtual reference station (VRS) method, partial derivative method, linear interpolation method and condition adjustment method. Among them, the virtual reference station technology is the most mature. The VRS system integrates GPS, Internet, wireless communication and computer network management. The whole system consists of several (more than 3) GPS fixed base stations, a GPS network control center and multiple end users.

③依据底图采集数据。

③ Data Collection based on base map.

为了充分利用已有的测绘成果，可通过数字化地形图的方法将纸质地形图转换为数字地形图，通常可使用跟踪数字化仪和扫描数字化仪。因跟踪数字化仪所得到的数字地图的精度低于原图，故已逐渐被扫描数字化仪所代替。

To make full use of the existing surveying and mapping results, paper topographic maps can be converted into digital topographic maps by digitizing topographic maps. The original map can be digitized usually with a tracking digitizer or scan digitizer. Since the digital map digitized with the tracking digitizer has a lower accuracy than the original map, it has been gradually replaced by the scan digitizer.

### （2）数字化测图的数据处理

(2) Data processing for digital surveying and mapping

数据处理主要是指采集数据结束至图形输出前的阶段，对各种图形数据的处理。数据处理包括数据传输、数据预处理、数据转换、数据计算、图形生成、图形编辑与整饰、图形信息的管理与使用等，它是数字化测图的关键。其中，数据转换是指将测量坐标转换为屏幕坐标。在测量坐标系中，坐标系原点在图幅的左下角，向上为 $x$ 轴正方向（北），向右为 $y$ 轴正方向（东）；而在计算机显示器中，坐标系原点在屏幕的左上角，向右为 $x$ 轴正方向，向下为 $y$ 轴正方向。因此，需要将测量坐标系的原点平移至屏幕左上角，并将测量坐标系按顺时针方向旋转 90°，这样可完成测量坐标向屏幕坐标的转换。

Data processing mainly refers to the processing of graphic data from the end of data collection to the stage before graphics output. Data processing, including data transmission, data preprocessing, data conversion, data calculation, graphic generation, graphic editing and finishing, and management and use of graphic information, is the key to digital mapping. Data conversion refers to the conversion of measured coordinates into screen coordinates. In the surveying coordinate system, the origin of the coordinate system is in the lower-left corner of the map, the vertical line pointing straight up is the positive direction of the $x$-axis (north), and the line pointing horizontally to the right is the positive direction of the $y$-axis (east). When the coordinate system is displayed in a computer, its origin is in the upper-left corner of the screen, the line pointing horizontally to the right is the positive direction of the $x$-axis and the vertical line pointing straight down is the positive direction of the $y$-axis. To convert the coordinate system into a screen-displayed one, we need to translate the origin of the measurement coordinate system to the upper-left corner of the screen and rotate the measurement coordinate system by 90° clockwise, so that the conversion of measurement coordinates to screen coordinates can be completed.

经过数据处理后，可产生平面图形数据文件和数字地面模型文件，然后对原图进行修改、编辑和整理，加上文字和高程注记，填充相应地物符号，再经过图形拼接、分幅和整饰，就可得到一幅规范的地形图。

After plane graphic data files and digital ground model files are generated through data processing, the original map can be modified, edited and sorted out. Then, we add text and elevation notes and fill in corresponding feature symbols to obtain a standard topographic map after graphics splicing, subdivision and finishing.

### （3）数字化测图的图形输出

(3) Graphic output for digital surveying and mapping

经过数据处理后得到的数字地形图是一个图形文件，它既可以永久地被保存在磁盘上，也可以转换成地理信息图形格式，用以建立或更新 GIS 图形数据库。图形输出是数字测图的主要目的，一般由计算机软件控制绘图仪自动绘出地形图。绘图仪的基本功能是实现 $x$、$y$ 坐标串和模（矢量）的转换，将计算机中数字图形描绘到图纸上。

The digital topographic map obtained after data processing is a graphic file, which is not only permanently saved on disk but also can be converted into the format of geographical information graphics to establish or update the GIS graphic database. Graphic output is the main purpose of digital mapping. Generally, computer software is used to control a plotter to automatically plot the topographic map. The basic function of the plotter is to realize the conversion of $x$, $y$ coordinate string and modulus (vector), and plot the digital graphics in the computer onto the map sheet.

# 本章小结

## Chapter Summary

本章主要讲述了地形图基本知识、大比例尺地形图测绘和地形图的应用等内容。

This chapter mainly introduces the basic knowledge of topographic maps, the mapping of large-scale topographic maps, and the application of topographic maps.

1. 地形图是通过实地测量，将地面上各种地物、地貌的平面位置和高程位置，按一定的比例尺，用统一规定的符号和注记，缩绘在图纸上的平面图形，它既表示地物的平面位置，又表示地貌形态。通常把 1 ： 500、1 ： 1 000、1 ： 2 000、1 ： 5 000 比例尺的地形图，称为大比例尺地形图。

1. A topographic map is a plane figure bearing plane positions and elevation positions of ground objects and land features on the ground that are scaled down at a scale and represented with uniform symbols and notes. It shows both the plane positions of ground objects and the land features. Topographic maps of scales 1 ： 500, 1 ： 1 000, 1 ： 2 000 and 1 ： 5 000 are usually referred to as large-scale topographic maps.

2. 大比例尺地形图测绘应重点掌握测图前的准备工作、碎部测量和地形图的绘制。碎部测量的方法有经纬仪测绘法、平板测图法等传统方法，也有全站仪测图法、GPS 网络 RTK 测图法等现代方法。

2. Surveying and mapping of large-scale topographic maps shall focus on the preparation before surveying and mapping, surveying of detail points and drawing of topographic maps. Surveying of detail points can be practiced with theodolites, plane tables and other traditional ways, as well as total stations, GPS network RTK and other modern ways.

3. 大比例尺地形图的应用主要包括地形图应用的基本内容和地形图在工程建设中的具体应用。地形图应用的基本内容包括确定图上某点的坐标、根据图上直线的长度确定水平距离、确定两点间直线的坐标方位角、确定点的高程、确定直线的坡度、图形面积的量算。地形图在工程建设中的应用包括按规定的坡度选择最短路线、按指定方向绘制纵断面图、确定汇水面积、根据地形图平整场地。

3. The application of large-scale topographic maps mainly includes the basic contents of topographic map application and the specific application of topographic maps in engineering construction. The basic contents of topographic map application include determining the coordinates of a certain point on the map, determining the horizontal distance according to the length of the corresponding straight line on the map, determining the coordinate and azimuth angle of the straight line through two points, determining the elevation of the point, determining the gradient of the straight line, and measuring the graphic area. The application of topographic maps in engineering construction includes selecting the shortest line based on a specified gradient, drawing the profile along a specified direction, determining the catchment area, and leveling the site based on topographic maps.

# 复习思考题

## Review Questions

### 一、填空题

### I. Fill in the blanks

1.______________是指地球表面上轮廓明显、具有固定性的物体，是指地球表面高低起伏的形态。

1.______________refers to a fixed object that has a clear contour on the earth's surface and refers to the undulating shape of the earth's surface.

2. 比例尺可分为______________和______________两种。

2. Scales can be categorized into two types:______________and______________.

3. 地形图的分幅方法分为______________和______________两类。

3. There are two methods to subdivide a topographic map:______________and______________.

4. 地形图的符号分为______________和______________，这些符号总称为______________。

4. Symbols of topographic map are categorized into______________and______________, and are collectively known as______________.

5. 大比例尺地形图测图前的准备工作主要有______________、______________和______________。

5. Preparations before surveying large-scale topographic maps mainly include__________, __________, and______________.

### 二、选择题

### II. Choice Questions

1. 地形图的比例尺用分子为 1 的分数形式表示时，（ ）。

1. When the scale of the topographic map is expressed as a fraction with the numerator of 1, ( ).

A. 分母越大，比例尺越大，表示地形越详细

A. the larger the denominator, the larger the scale, and the more detailed the terrain

B. 分母越小，比例尺越小，表示地形越概略

B. the smaller the denominator, the smaller the scale, and the briefer the terrain

C. 分母越大，比例尺越小，表示地形越详细

C. the larger the denominator, the smaller the scale, and the more detailed the terrain

D. 分母越小，比例尺越大，表示地形越详细

D. the smaller the denominator, the larger the scale, and the more detailed the terrain

2. 地形图上（ ）mm 所代表的实地水平距离，称为比例尺精度。

2. The scale accuracy refers to the real-world horizontal distance represented by ( ) mm on the topographic map.

A. 0.1 B. 0.2 C. 0.3 D. 0.4

3. 下列四种比例尺地形图，比例尺最大的是（ ）。

3. Among the following four scale topographic maps, the largest scale is ( ).

A. 1:1 0000 B. 1:5 000 C. 1:2 000 D. 1:1 000

4. 山脊线也称（ ）。

4. The ridge line is also known as ( ).

A. 示坡线 B. 分水线 C. 山谷线 D. 集水线

A. slope line B. watershed line C. valley line D. watercourse line

5. 下列不属于等高线特征的是（ ）。

5. Which of the following is not the characteristic of contour lines? ( )

A. 同一条等高线上各点的高程必相等

A. All points on a contour line are of the same elevation.

B. 等高线是一闭合曲线，如不在本幅图内闭合，则在相邻的其他图幅内闭合

B. All contour lines must close either within the map boundary or outside.

C. 不同高程的等高线不能相交或重合

C. Contour lines of different elevations cannot cross each other or unite to form one line.

D. 在同一幅图内，等高线的平距大，表示地面坡度缓；平距小，则表示地面坡度陡；平距相等，则表示坡度相同

D. On the same topographic map, the larger the horizontal distance, the gentler the ground slope is; conversely, the steeper it is; equal horizontal distance indicates same gradient.

## 三、简答题

## III. Short Answer Questions

1. 什么是地形图比例尺？地形图比例尺可分为哪两种？

1. What is the scale of a topographic map? What are the two types of scales of topographic maps?

2. 地形图中地物、地貌如何表示？

2. How to represent ground objects and land features on a topographic map?

3. 什么是等高线、等高线平距与等高距？等高线的分类及特性有哪些？

3. What are the contour line, the horizontal distance of contour lines and the contour interval? What are

the types and characteristics of contours?

4. 典型地貌有哪些？其等高线各有什么特点？试绘图说明。

4. What are the typical land features? What are the characteristics of their contour lines? Please illustrate them with drawings.

5. 地形图测绘前的准备工作有哪些？

5. What are the preparations before surveying and mapping topographic maps?

6. 如何进行地形图的绘制？

6. How to draw a topographic map?

6

建筑工程测量

ARCHITECTURAL ENGINEERING SURVEYING

# 民用建筑施工测量

# Construction Survey of Civil Buildings

**学习目标：**

**Learning Objectives:**

通过本章的学习，了解民用建筑施工测量前的准备工作；熟悉民用建筑的定位与放线、基础工程施工测量方法，以及墙体工程施工测量中各层轴线测设及标高引测方法；掌握高层建筑施工测量中的竖向测量方法以及高层建筑的高程传递的方法。

Through the study of this chapter, to learn about how to prepare for construction survey of civil buildings; to get familiar with the positioning and setting-out of civil buildings, foundation engineering construction survey methods, and axis projection and elevation piloting survey of each floor in wall works; to master the vertical survey methods and elevation transfer of high-rise buildings.

**能力目标：**

**Ability Objectives:**

具备民用建筑物定位与放线及施工测量的能力，能够进行高层建筑的竖向测量与高程传递测设。

Be able to conduct positioning, setting-out and construction survey of civil buildings, and carry out vertical survey and elevation transfer survey of high-rise buildings.

民用建筑是指住宅、办公楼、商场、医院、宾馆、学校等建筑物。民用建筑施工测量的任务是按照设计的要求，把民用建筑的位置（平面位置和高程）测设到地面上，并配合施工以保证工程质量。因此，民用建筑施工测量的主要任务是建筑物的定位和放线、基础工程施工测量、墙体工程施工测量及高层建筑施工测量等。

Civil buildings refer to residential buildings, office buildings, shopping malls, hospitals, hotels, schools, etc. When it comes to construction survey of civil buildings, you need to set out the positions (plane positions and elevations) of civil buildings on the ground according to the design requirements, and coordinate with the construction to ensure the project quality. In this regard, the main tasks of civil building construction survey include building positioning and setting-out, construction survey for foundation works, construction survey for wall works and construction survey for high-rise buildings.

# 6.1 施工测量前的准备工作

# 6.1 Preparation before Construction Survey

## 6.1.1 熟悉设计图纸

## 6.1.1 Get Familiar with Design Drawings

设计图纸是施工测量的主要依据，在测设前，应熟悉建筑物的设计图纸，了解施工建筑物与相邻地物的相互关系，以及建筑物的尺寸和施工的要求等，并仔细核对各设计图纸的有关尺寸。与测设工作有关的设计图纸主要有建筑总平面图、建筑平面图、基础平面图、基础详图和建筑物的立面图及剖面图。

The design drawings serve as the main basis for construction survey, so you should get familiar with the design drawings of the buildings before setting out, understand the relationship between the construction buildings and adjacent ground objects, and the sizes and construction requirements of the buildings, and carefully check the sizes on each design drawing. The design drawings involved in setting-out mainly include general layout plan, building plan, foundation plan, foundation details, and elevation and section of a building.

①建筑总平面图（图 6-1）。从总平面图上，可以查取或计算设计建筑物与原有建筑物或测量控制点之间的平面尺寸和高差，作为测设建筑物总体位置的依据。

① General Layout Plan of Building (Figure 6-1). On the general layout plan, you can calculate and obtain the plane size and elevation difference between the design building and the existing buildings or survey control points, as the basis for setting out the general location of the building.

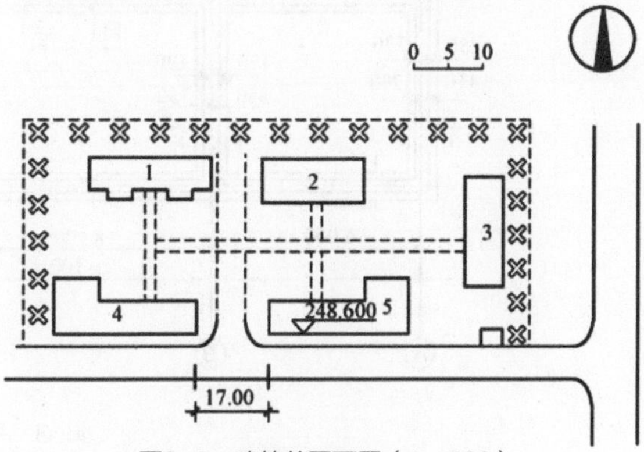


图6-1 建筑总平面图（1：500）

Figure 6-1 General Layout of Building (1：500)

②建筑平面图(图6-2)。从建筑平面图中，可以查取建筑物的总尺寸，以及内部各定位轴线之间的关系尺寸。它是测设建筑物细部轴线的依据，是施工测设的基本资料，要注意其尺寸是否与建筑总平面图的尺寸相符。

② Building Plan (Figure 6-2). From the building plan, you can obtain the overall dimensions of the building and the dimension relationship between the internal positioning axes. It is not only the basis for setting out the detailed axes of the building, but also the basic document for construction survey. So, you need to check whether the sizes are consistent with the general layout plan of the building.

③基础平面图(图6-3)。从基础平面图中，可以查取基础边线与定位轴线的平面尺寸，这是测设基础轴线的必要数据。

③ Foundation Plan (Figure 6-3). On the foundation plan, you can check the plane sizes of the foundation boundaries and the positioning axes, which are the necessary data for setting out the foundation axes.

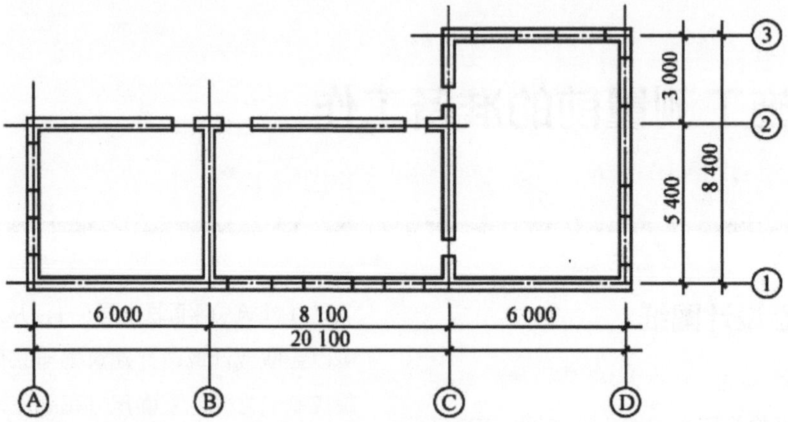


图6-2 建筑平面图

Figure 6-2 Building Plan

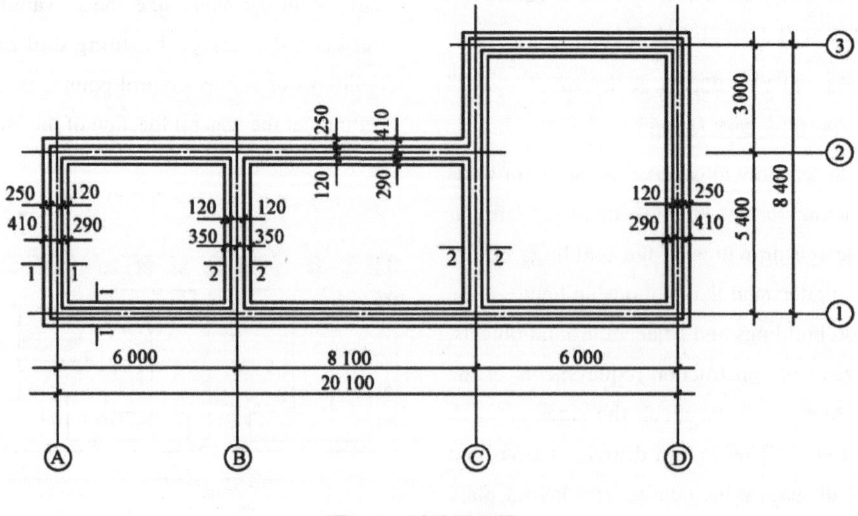


图6-3 基础平面图

Figure 6-3 Foundation Plan

④基础详图(图6-4)。从基础详图中，可以查取基础立面尺寸和设计标高，这是基础高程测设的依据。

④ Foundation Details (Figure 6-4). On the foundation details, you can find the elevation sizes and design elevations of the foundation, which provide a basis for the foundation elevation survey.

⑤建筑物的立面图和剖面图。从建筑物的立面图和剖面图中，可以查取基础、地坪、门窗、楼板、屋架和屋面等设计高程，这是高程测设的主要依据。

⑤ Elevations and sections of buildings. On the elevations and sections of a building, you can find the design elevations of foundations, floors, doors and windows, floor slabs, roof trusses and roofs, which provide the main basis for elevation survey.

## 6.1.2 现场踏勘

### 6.1.2 Site Survey

全面了解施工现场情况，检核测设仪器和工具，对施工场地上的平面控制点和水准点进行检核，平整和清理施工场地，以便进行测设工作，根据设计要求、定位条件、现场地形和施工方案等因素制定测设方案，包括测设方法、测设数据计算和绘制测设略图。

It is necessary to comprehensively understand the conditions of the construction site, check the instruments and tools for setting-out and the plane control points and benchmarks on the site, level and clean the construction site for setting-out, and work out related schemes according to the design requirements, positioning conditions, site topography, construction schemes, etc., covering setting-out methods, calculation of setting-out data and drawing of setting-out sketch.

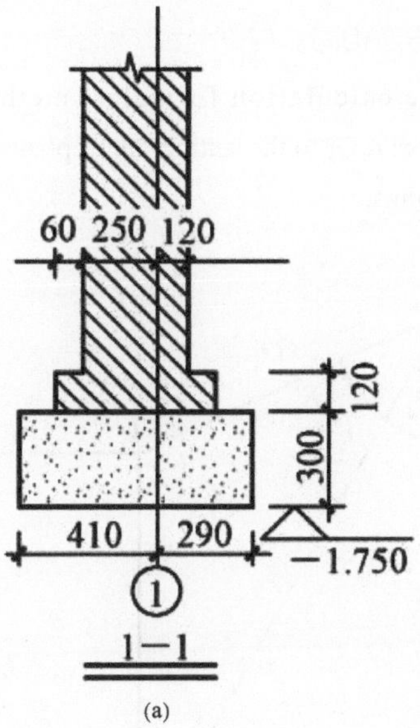

(a)

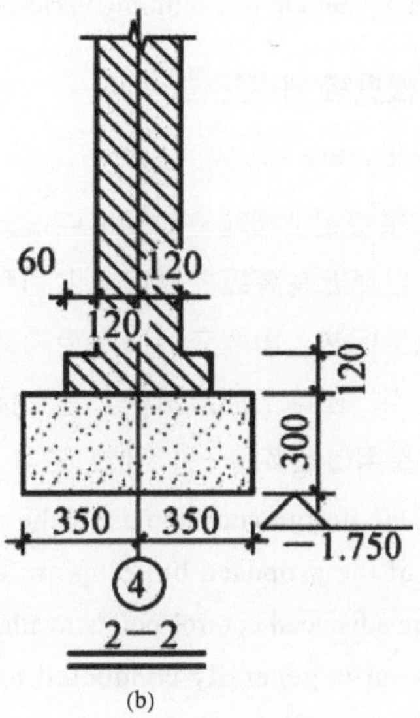

(b)

图6-4 基础详图

Figure 6-4 Foundation Details

# 6.2 建筑物的定位和放线

## 6.2 Positioning and Setting-out of Buildings

### 6.2.1 建筑物的定位

### 6.2.1 Positioning of Buildings

建筑物的定位，就是把建筑物外廓各轴线的交点测设在地面上，然后再根据这些点进行细部放样和基础放线。

The building positioning is just to set out the intersection points of axes along the building external contour lines on the ground, and then carry out detailed setting-out and foundation setting-out according to these points.

根据现场条件的不同，建筑物的定位方法也有所不同。

Depending on the site conditions, the positioning method of a building varies.

#### （1）根据控制点定位

#### (1) Positioning based on control points

如果待定位建筑物的定位点设计坐标是已知的，且附近有高级控制点可供利用，一般都选用极坐标法、角度交会法或距离交会法来测设定位点。在这三种方法中，极坐标法适用性最强，是用得最多的一种定位方法。

If the design coordinates of the positioning points of the proposed building are known and there are advanced control points available nearby, setting out is generally conducted by the polar coordinate method, angular intersection method or distance intersection method. Among the three methods, the polar coordinate method is the most applicable and the most commonly used.

如图 6-5 所示，测量控制点 $A$、$B$ 及拟建建筑物外角点 1、2 坐标由设计图纸给定。若 1 点用极坐标法测设，则要计算出图中的角 $\beta_1$ 及距离 $D_1$。若 2 点用角度交会法测设，利用相应点的坐标反算出各边的方位角，就可计算夹角 $\alpha_2$ 及 $\beta_2$。

As shown in Figure 6-5, the coordinates of survey control points $A$ and $B$ and the outer corner points 1 and 2 of the proposed building are given in the design drawing. If point 1 is set out using the polar coordinate method, the angle $\beta_1$ and distance $D_1$ in the figure shall be calculated. If point 2 is set out using the angular intersection method, the azimuth angles of each side can be calculated using the coordinates of corresponding points, and then the intersection angles $\alpha_2$ and $\beta_2$ can be calculated.

计算公式、方法及测设过程与点的平面位置测设方法相同。

The calculation formulas, methods and procedures refer to the setting-out of plane positions of the points.

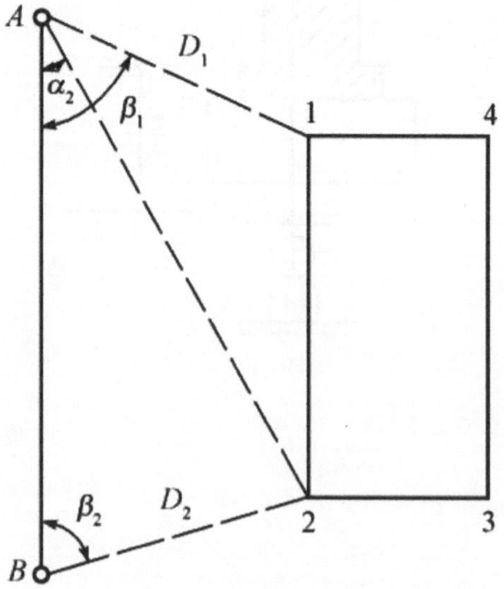


图6-5 根据控制点定位

Figure 6-5 Positioning Based on Control Points

## （2）根据建筑方格网和建筑基线定位

## (2) Positioning based on building square grids and building baselines

如果待定位建筑物的定位点设计坐标是已知的，且建筑场地已设有建筑方格网或建筑基线，可利用直角坐标法测设定位点。用直角坐标法测设点位，所需测设数据的计算较为方便，在用经纬仪和钢尺实地测设时，建筑物总尺寸和四大角的精度容易控制和检核。

If the design coordinates of the positioning points of the proposed building are known and there are building grids or building baselines available on the building site, the positioning points can be set out and surveyed by rectangular coordinate method. In fact, the rectangular coordinate method provides more convenience in calculation of required data, while the accuracy of the overall dimensions and four major corners of the building is easy to be controlled and checked after using a theodolite and steel ruler.

## （3）根据与原有建筑物和道路的关系定位

## (3) Positioning based on the relationship with existing buildings and roads

如果设计图上只给出新建建筑物与附近原有建筑物或道路的相互关系，而没有提供建筑物定位点的坐标，周围又没有测量控制点、建筑方格网和建筑基线可供利用，可根据原有建筑物的边线或道路中心线，将新建建筑物的定位点测设出来。

If, as shown in the design drawing, there is only the relationship between the new building and the existing building or road nearby, but no coordinates of the building positioning points, and there are no survey control points, building square grids and building baselines available, the positioning points of the new building can be set out according to the boundaries of the existing building or the road centerline.

具体测设方法因实际情况的不同而不同，但基本过程是一致的：在现场先找出原有建筑物的边线或道路中心线，再用经纬仪和钢尺将其延长、平移、旋转或相交，得到新建建筑物的一条定位轴线，然后根据这条定位轴线，用经纬仪测设角度（一般是直角），用钢尺测设长度，得到其他定位轴线或定位点，最后检核四个大角和四条定位轴线长度是否与设计值一致。下面分两种情况说明具体测设方法。

Specifically, the setting-out method varies depending on the actual situation, but the basic process is the same, that is, we should first find out the boundaries of the existing building or the centerline of the road on site, then extend, translate, rotate or intersect them using a theodolite and a steel ruler to obtain a positioning axis of the new building; next, measure the angle (generally a right angle) with a theodolite and the length with a steel ruler based on this positioning axis to obtain other positioning axes or positioning points, and finally check whether the four major corners and the lengths of the four positioning axes are consistent with the design values. The specific setting-out methods are described in the following two cases.

①根据与原有建筑物的关系定位。如图 6-6(a) 所示，拟建建筑物的外墙边线与原有建筑物的外墙边线在同一条直线上，两栋建筑物的间距为 10 m，拟建建筑物四周长轴为 40 m，短轴为 18 m，轴线与外墙边线间距为 0.12 m，可按下述方法测设其四个轴线交点。

① Positioning based on the relationship with the existing building. As shown in Figure 6-6 (a), the outer wall boundary lines of the proposed building and the existing building are on the same straight line; the spacing between the two buildings is 10 m; the major axis of the proposed building is 40 m, the minor axis is 18 m, and the spacing between the axis and the outer wall boundary line is 0.12 m. The intersection of the four axes can be set out as follows.

a. 沿原有建筑物的两侧外墙拉线，用钢尺顺线从墙角往外量一段较短的距离（这里设为 2 m），在地面上定出 $T_1$ 和 $T_2$ 两个点，$T_1$ 和 $T_2$ 的连线即为原有建筑物的平行线。

a.Set out line along both sides of the outer wall of the existing building, measure a short distance from the corner of the wall with a steel ruler (2 m tentatively), and set points $T_1$ and $T_2$ on the ground. The connecting line between $T_1$ and $T_2$ is the parallel line to the existing building.

b. 在 $T_1$ 点安置经纬仪，照准 $T_2$ 点，用钢尺从 $T_2$ 点沿视线方向量 10 m + 0.12 m，在地面上定出 $T_3$ 点，再从 $T_3$ 点沿视线方向量 40 m，在地面上定出 $T_4$ 点，$T_3$ 和 $T_4$ 的连线即为拟建建筑物的平行线，其长度等于长轴尺寸。

b.Place a theodolite at point $T_1$ and aim at point $T_2$ , measure 10 m+0.12 m from point $T_2$ along the line of sight using a steel ruler, set point $T_3$ on the ground, then measure 40 m from point $T_3$ along the line of sight, and set point $T_4$ on the ground. The connecting line between $T_3$ and $T_4$ is the parallel line of the proposed building, with a length equal to the major axis.

c. 在 $T_3$ 点安置经纬仪，照准 $T_4$ 点，逆时针测设 90°，在视线方向上量 2 m + 0.12 m，在地面上定出 $P_1$ 点，再从 $P_1$ 点沿视线方向量 18 m，在地面上定出 $P_4$ 点。同理，在 $T_4$ 点安置经纬仪，照准 $T_3$ 点，逆时针测设 90°，在视线方向上量 2 m + 0.12 m，在地面上定出 $P_2$ 点，再从 $P_2$ 点沿视线方向量 18 m，在地面上定出 $P_3$ 点。则 $P_1$、$P_2$、$P_3$ 和 $P_4$ 点即为拟建建筑物的四个定位轴线点。

c.Place a theodolite at point $T_3$ and aim at point $T_4$, measure 90° counter-clockwise, measure 2 m+0.12 m in the line-of-sight direction, set point $P_1$ on the ground, then measure 18 m along the line-of-sight direction from point $P_1$, and set point $P_4$ on the ground. Similarly, place a theodolite at point $T_4$ and aim at point $T_3$, measure 90° counter-clockwise, measure 2 m+0.12 m in the line-of-sight direction, set point $P_2$ on the ground, and then measure 18 m along the line-of-sight direction from point $P_2$ and set point $P_3$ on the ground. Points $P_1$, $P_2$, $P_3$ and $P_4$ are the four positioning axis points of the proposed building.

d. 在 $P_1$、$P_2$、$P_3$ 和 $P_4$ 点上安置经纬仪，检核四个大角是否为 90°，限差在 ±40″ 范围内。用钢尺丈量四条轴线的长度，检核长轴是否为 40 m，短轴是否为 18 m，相对误差不超过 1/3 000 即为合格。

d.Place theodolites at points $P_1$, $P_2$, $P_3$ and $P_4$ to check whether the four corners are 90°, and the tolerance is within ±40″. Measure the length of the four axes with a steel ruler, and check whether the major axis is 40 m and the minor axis is 18 m. It is acceptable if the relative error is not more than 1/3 000.

如果是如图 6-6(b) 所示的情况，则在得到原有建筑物的平行线并延长到 $T_3$ 点后，在 $T_3$ 点测设 90° 并量距，定出 $P_1$ 和 $P_2$ 点，得到拟建建筑物的一条长轴，再分别在 $P_1$ 和 $P_2$ 点测设 90° 并量距，定出另一条长轴上的 $P_4$ 和 $P_3$ 点。注意不能先定短轴的两个点（如 $P_1$ 和 $P_4$ 点），再在这两个点上设站测设另一条短轴上的两个点（如 $P_2$ 和 $P_3$ 点），否则误差容易超限。

If it is as shown in Figure 6-6(b), after the parallel line to the existing building is obtained and extended to point $T_3$, a 90° angle is measured at point $T_3$ and the distance is measured to set points $P_1$ and $P_2$ to obtain a major axis of the proposed building; then, 90° is measured at points $P_1$ and $P_2$ respectively while the distance is measured to set points $P_4$ and $P_3$ on the other major axis. Note that two points (e.g. $P_1$ and $P_4$) on the minor axis cannot be set first, and then two points (e.g. $P_2$ and $P_3$) on the other minor axis, otherwise the error is prone to exceed the limited value.

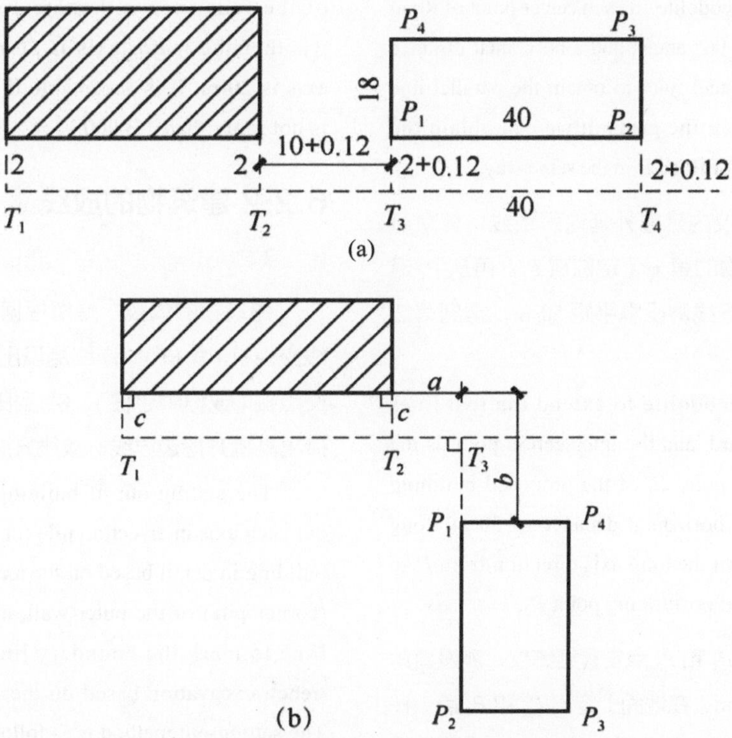


图6-6　根据与原有建筑物的关系定位

Figure 6-6 Positioning Based on the Relationship with the Existing Building

②根据与原有道路的关系定位。如图 6-7 所示，拟建建筑物的轴线与道路中心线平行，轴线与道路中心线的距离如图 6-7 所示，测设方法如下。

② Positioning based on the relationship with the existing road. As shown in Figure 6-7, the axis of the proposed building is parallel to the road centerline, and the distance between the axis and the road centerline is shown in Figure 6-7. The setting-out method is as follows.

a. 在每条道路上选两个合适的位置，分别用钢尺测量该处道路宽度，其宽度的 1/2 处即为道路中心点，如此得到路 I 中心线的两个点 $C_1$ 和 $C_2$，同理得到路 II 中心线的两个点 $C_3$ 和 $C_4$。

a.Choose two suitable points on each road and measure the road width with a steel ruler. The road center point is just at the 1/2 of the width. In this way, we can obtain two points $C_1$ and $C_2$ on the center line of Road I, and $C_3$ and $C_4$ on the center line of Road II.

b. 分别在路 I 的两个中心点上安置经纬仪，测设 90°，用钢尺测设水平距离 12 m，在地面上得到路 I 的平行线 $T_1T_2$，同理作出路 II 的平行线 $T_3T_4$。

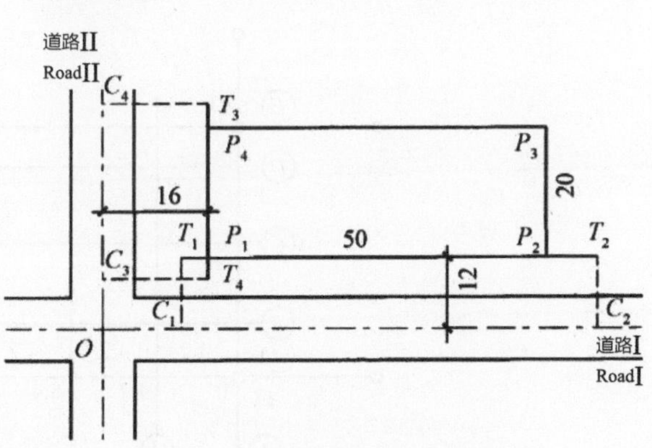


图6-7　根据与原有道路的关系定位

Figure 6-7 Positioning Based on the Relationship with the Existing Road

b.Place a theodolite on each center point of Road I, and measure a 90° angle, and a horizontal distance of 12 m with a steel ruler to obtain the parallel line $T_1T_2$ of Road I on the ground; we can obtain the parallel line $T_3T_4$ of Road II in the same way.

c. 用经纬仪内延或外延这两条线，其交点即为拟建建筑物的第一个定位点 $P_1$，再从 $P_1$ 沿长轴方向的平行线测设水平距 50 m，得到第二个定位点 $P_2$。

c.Use a theodolite to extend the two lines inward or outward, and the intersection point is the first positioning point $P_1$ of the proposed building; then, measure a horizontal distance of 50 m along the parallel line in the long axis direction from $P_1$ to obtain the second positioning point $P_2$.

d. 分别在 $P_1$ 和 $P_2$ 点安置经纬仪，测设直角和水平距离 20 m，在地面上定出 $P_3$ 和 $P_4$ 点。在 $P_1$、$P_2$、$P_3$ 和 $P_4$ 点上安置经纬仪，检核角度是否为 90°，限差在 ±40″ 以内。用钢尺丈量四条轴线的长度，检核长轴是否为 50 m，短轴是否为 20 m，相对误差不超过 1/3 000 即为合格。

d.Place a theodolite at points $P_1$ and $P_2$ respectively, measure a right angle and horizontal distance of 20 m, so as to set points $P_3$ and $P_4$ on the ground. Place theodolites at points $P_1$, $P_2$, $P_3$ and $P_4$ to check whether the corners are 90°, and the tolerance is within ±40″. Measure the length of the four axes with a steel ruler, and check whether the major axis is 50 m and the minor axis is 20 m. It is acceptable if the relative error is not more than 1/3 000.

## 6.2.2 建筑物的放线

### 6.2.2 Setting-out of Building

建筑物的放线，是指根据已定位的外墙轴线交点桩（角桩），详细测设出建筑物各轴线的交点桩（或称中心桩），然后根据交点桩用白灰撒出基槽开挖边界线。放线方法如下。

The setting-out of building refers to setting out each axis intersection pile (or center pile) of the building in detail based on the axis intersection pile (corner pile) of the outer wall, and then spreading lime to mark the boundary line for foundation trench excavation based on the intersection piles. The setting-out method is as follows.

### （1）测设细部轴线交点

#### (1) Set out the detailed axis intersections

如图 6-8 所示，*A* 轴、*E* 轴、①轴和⑦轴是建筑物的四条外墙主轴线，其交点 *A*1、*A*7、*E*1 和 *E*7 是建筑物的定位点，这些定位点已在地面上测设完毕并打好桩点，各主次轴线间隔如图 6-8 所示，现欲测设次要轴线与主轴线的交点。

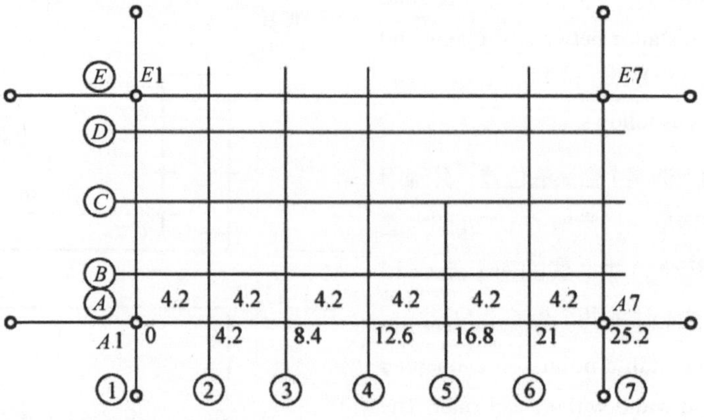


图6-8 测设细部轴线交点

Figure 6-8 Set Out the Detailed Axis Intersections

As shown in Figure 6-8, axes *A*, *E*, ① and ⑦ are the four main axes of the outer wall of the building, and their intersection points *A*1, *A*7, *E*1 and *E*7 are the positioning points of the building. These positioning points have been surveyed and set on the ground and piled. The spacing between the main and secondary axes is shown in Figure 6-8. The intersection points of the secondary axes and the main axes are now to be surveyed and set out.

在 *A*1 点安置经纬仪，照准 *A*7 点，把钢尺的零端对准 *A*1 点，沿视线方向拉钢尺，在钢尺上读数等于①轴和②轴间距 (4.2 m) 的地方打下木桩，打的过程中要经常用仪器检查桩顶是否偏离视线方向，并不时拉一下钢尺，看钢尺读数是否还在桩顶上，如有偏移要及时调整。打好桩后，用经纬仪视线指挥在桩顶上画一条纵线，再拉好钢尺，在读数等于轴间距处画一条横线，两线交点即 *A* 轴与②轴的交点 *A*2。

Place a theodolite at point *A*l and aim at point *A*7, align the zero end of the steel ruler with point *A*l, pull the steel ruler along the line-of-sight direction, and drive a wooden pile where the steel ruler reading is equal to the spacing between axes ① and ② (4.2 m). During driving, check whether the pile top deviates from the line of sight with an instrument, and pull the steel ruler from time to time to see whether the reading is still on the pile top. Any deviation shall be adjusted immediately. After the pile is driven, draw a longitudinal line on the top of the pile with a theodolite, and then use a steel ruler to draw a horizontal line where the reading is equal to the axis distance, so the intersection point of the two lines is the intersection point *A*2 of axes *A* and ②.

在测设 *A* 轴与③轴的交点 *A*3 时，方法同上，注意仍然要将钢尺的零端对准 *A*1 点，沿视线方向拉钢尺，而钢尺读数应为①轴和③轴间距 (8.4 m)，这种做法可以减小钢尺对点误差，避免轴线总长度增长或减短。如此依次测 *A* 轴与其他有关轴线的交点。测设完最后一个交点后，用钢尺检查各相邻轴线桩的间距是否等于设计值，误差应小于 1/3 000。

Follow the above steps for setting out the intersection point *A*3 of axes *A* and ③. Note that the zero end of the steel ruler shall still be aligned with *A*l, and the steel ruler shall be pulled along the line-of-sight direction, and the reading of the steel ruler shall be the spacing between the axis ① and the axis ③ (8.4 m). In this way, it allows to reduce the point-to-point error of the steel ruler and prevent the increase or decrease in the total length of the axis. Similarly, measure the intersection points of axis *A* and other relevant axes. After the last intersection point is surveyed and set out, check whether the spacing between adjacent axis piles is equal to the design value with a steel ruler, and the error shall be less than 1/3 000.

测设完 *A* 轴上的轴线点后，用同样的方法测 *E* 轴、①轴和⑦轴上的轴线点。如果建筑物的尺寸较小，也可用拉细线绳的方法代替经纬仪定线，然后沿细线绳拉钢尺量距。此时要注意细线绳不要碰到物体，风大时也不宜作业。

After the axis points on axis *A* is done, measure the axis points on axis *E*, ① and ⑦ in the same way. If the building has small sizes, you can locate a route by a string instead of theodolite, and then measure the distance with a steel ruler along the string. If so, pay attention not to touch the string, and do not work in the windy days.

### （2）引测轴线

### (2) Axis transfer survey

在基槽或基坑开挖时，定位桩和细部轴线桩均会被挖掉，为了使开挖后各阶段施工能准确地恢复各轴线位置，应把各轴线延长到开挖范围以外的地方并做好标志，这个工作称为引测轴线，具体有设置龙门板和轴线控制桩两种形式。

During the excavation of foundation trench

or foundation pit, the positioning piles and detailed axis piles will be excavated. In order to restore the normal position of each axis accurately at each stage of construction after excavation, each axis shall be extended beyond the scope of excavation and marked. This work refers to axis transfer survey, and there are two ways: batter board and axis control peg.

①设置龙门板。

① Setting of batter board.

在小型民用建筑中，常将各轴线引测到基槽外的水平木板上，水平木板称为龙门板，固定龙门板的木桩称为龙门桩。如图 6-9 所示，在建筑物四角与隔墙两端及槽开挖边界线以外 1.5~2.0 m 处设置龙门桩，龙门桩要钉得竖直和牢固，龙门桩的外侧应与基槽平行。利用施工场地的水准点，将建筑物的室内地坪设计标高 (±0.000) 引测到龙门桩的外侧团上，用红铅笔画一横线。沿龙门桩上 ±0.000 标高处钉设龙门板。如果施工现场地面太高或太低，也可测设比 ±0.000 高或低一整数的标高线钉设龙门板，并注明。用水准仪校核龙门板的高程，如有差错应及时纠正，其允许误差为 ±5 mm。

For small civil buildings, each axis is often transferred to a horizontal board outside the foundation trench. The horizontal board is called a batter board, and the wooden pile fixing the batter board is called a gantry pile. As shown in Figure 6-9, gantry piles shall be set at the four corners of the building, both ends of the partition wall and 1.5~2.0 m outside the trench excavation boundary line. The gantry piles shall be nailed vertically and firmly, and the outer side of the gantry piles shall be parallel to the foundation trench. Use the benchmark of the construction site to transfer the indoor floor design elevation (±0.000) of the building to the outer side of the gantry pile, and draw a horizontal line with a red pencil. The batter board is nailed along the gantry pile with an elevation of ±0.000. If the ground of the construction site is too high or too low, the batter board can be nailed with an elevation of an integer higher or lower than ±0.000, but it shall be indicated. Check the elevation of the batter board with a level gauge, and correct any error in time, with an allowable error of ±5 mm.

将经纬仪安置于轴线交点上，照准另一轴线交点后，将轴线引测到龙门板上，钉上小钉（称为中心钉）作为标志。用钢尺沿龙门板顶面实量各中心钉（轴线）间的距离是否正确，中心钉定位误差应小于 ±5 mm。经检核无误后，以中心钉为准，将墙边线、基础边线、基础开挖边线等标定在龙门板上。

Place a theodolite on the axis intersection

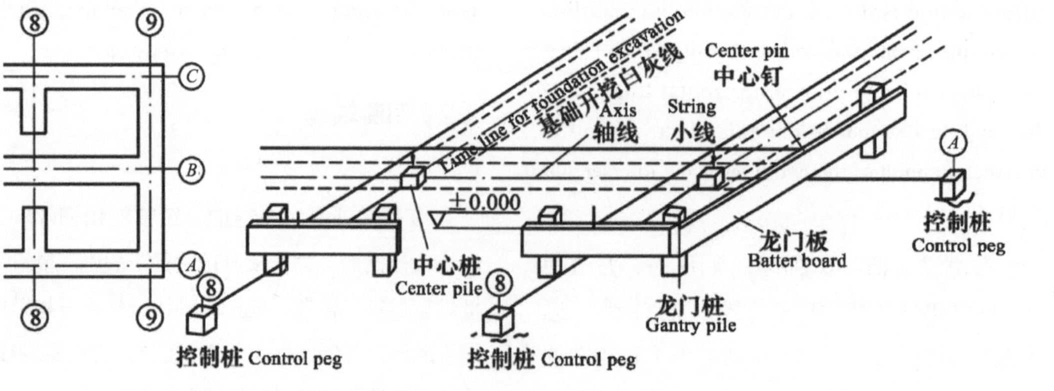


图6-9 龙门桩与龙门板

Figure 6-9 Gantry Pile and Batter Board

point and aim at another axis intersection point, transfer the axis to the batter board, and drive in a nail (center nail) as a mark. Use a steel ruler to measure whether the distance between the center nails (axes) along the top surface of the batter board is correct, and the positioning error of the center nails shall be less than ±5 mm. After verification, the wall boundary line, foundation boundary line and foundation excavation boundary line shall be marked on the batter board based on the center nail.

②设置轴线控制桩。

② Setting of axis control peg.

龙门板需要较多木料，而且占用场地，使用机械开挖时容易被破坏，因此也可以在基槽或基坑外各轴线的延长线上测设轴线控制桩，作为以后恢复轴线的依据。即使设置了龙门板，为了防止龙门板被碰撞而挪动位置，对主要轴线也应测设轴线控制桩。

Since there is more timber needed for the batter board, which will occupy a large area on the site and be prone to be damaged during mechanical excavation, the axis control peg can be surveyed and set on the extension line of each axis outside the foundation trench or foundation pit to provide a reference for future restoration of the axis. Even if a batter board is set, axis control pegs shall be set for the main axis to prevent the batter board from being collided and moved.

轴线控制桩一般设在开挖边线 4 m 以外的地方，并用水泥砂浆加固。最好是附近有固定建筑物和构筑物，这时应将轴线投测在这些物体上，用红漆标记，使轴线更容易得到保护，但每条轴线至少应有一个控制桩是设在地面上的，以便今后能通过安置经纬仪来恢复轴线。

Axis control pegs are generally located 4 m away from the excavation boundary line and reinforced with cement mortar. It is best that there are fixed buildings and structures nearby. At this time, the axis shall be transferred onto these objects and marked with red paint for protection. However, at least one control peg shall be set on the ground for each axis so that the theodolite can be placed to restore the axis in the future.

轴线控制桩的引测主要采用经纬仪法，当引测到较远的地方时，要注意采用盘左和盘右两次投测取中法来引测，以减少引测误差和避免出现错误。

The transfer survey of axis control pegs is mainly conducted using a theodolite. When it is transferred to a far place, the transfer survey shall be conducted from the face-left and face-right positions and then the median shall be taken to reduce the survey errors.

③用石灰撒出基槽开挖边线。

③ Liming to mark the boundary line for foundation trench excavation.

在轴线的两端，根据龙门板上标定出的基槽上口宽度拉直细线绳，并沿此线绳撒出白灰线，作为挖槽的依据。

Straighten the string at both ends of the axis according to the upper opening width of the foundation trench marked on the batter board, and then spread lime along this string as a reference for trench excavation.

一般建筑放线时，± 0.000 标高测设误差范围不大于 ± 5 mm，轴线间距离校核相对误差不得大于 1/3 000，其夹角允许误差范围不大于 ± 40"。

Generally, during building setting-out, the error range of ±0.000 elevation survey shall not be greater than ±5 mm, the relative error of axis distance shall not be greater than 1/3 000, and the allowable error range of intersection angle shall not be greater than ±40".

# 6.3 建筑物基础工程施工测量

# 6.3 Construction Survey of Building Foundation Works

## 6.3.1 水平桩的测设

## 6.3.1 Setting-out of Horizontal Piles

一般民用建筑大多采用条形基础。当基槽开挖到一定深度时，在基槽壁上自拐角开始，每隔 3~4 m，由龙门板上沿的 ±0.000 标高测设一比槽底设计标高高 0.5 m 的水平桩，作为挖槽深度、找平槽底和垫层的标高依据。一般在基槽各拐角处均应打水平桩，然后拉上白线，线下 0.5 m 即为槽底设计高程。水平桩可以是木桩也可以是竹桩。

Generally, strip foundations are adopted for most civil buildings. When the foundation trench is excavated to a certain depth, a horizontal pile 0.5 m higher than the design elevation of the trench bottom shall be surveyed and set out on the foundation trench wall from the corner every 3~4 m from ±0.000 of the upper edge of the batter board as a reference for trench excavation depth, leveling of the trench bottom and cushion elevation. Generally, horizontal piles shall be driven at each corner of the foundation trench, and then a white line shall be drawn, with 0.5 m below the line as the design elevation of the trench bottom. Horizontal piles can be either wood or bamboo.

如图 6-10 所示，室内地坪 (±0.000 m) 的设计标高为 49.800 m，槽底设计标高为 48.100 m，欲测设比槽底设计标高高 0.5 m 的水平桩。其标高为 48.100 + 0.5=48.600 m。在槽边适当处安置水准仪，在龙门板上立水准尺，读得后视读数为 0.774 m，则视线高为 49.800 + 0.774=50.574 m。求得水准尺立在水平桩上的前视读数为 50.574−48.600=1.974 m。在槽内一侧立水准尺，上下移动，当水准仪读数为 1.974 m 时，用一木桩水平地紧贴尺底钉入槽壁，即为所测的水平桩。同理，测设出其余各桩。水平桩测设的标高容许误差应在 ±10 mm 范围内。

As shown in Figure 6-10, the design elevation of the indoor floor (±0.000 m) is 49.800 m, and the design elevation of the trench bottom is 48.100 m. It is proposed to set a horizontal pile 0.5 m higher than the design elevation of the trench bottom. The elevation thereof is 48.100+0.5=48.600 m. A level gauge is placed at an appropriate position on the edge of the trench, and a leveling rod on the batter board, and the backsight reading is 0.774 m, so the elevation of sight is 49.800+0.774=50.574 m. The forward reading of the leveling rod on the horizontal pile is 50.574−48.600=1.974 m. A leveling rod is erected on one side of the trench and moved up and down. When the reading is 1.974 m,

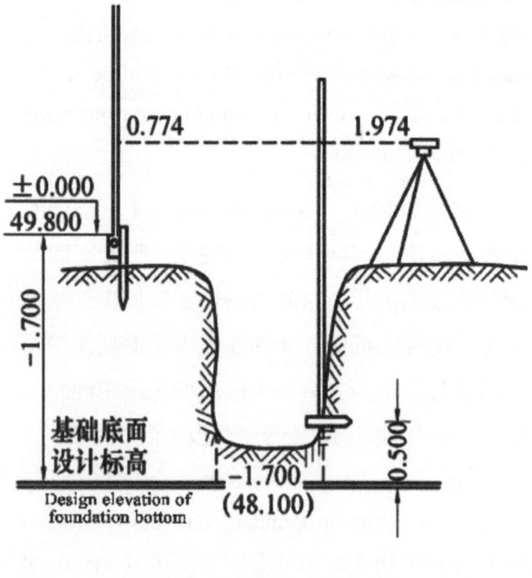


图6-10　基槽水平桩设置

Figure 6-10 Setting of Horizontal Pile in Foundation Trench

a wood pile is horizontally nailed into the trench wall against the bottom of the rod, which is just the horizontal pile. Other piles are surveyed and set out in the same way. The allowable error of elevation of horizontal pile shall be within ±10 mm.

## 6.3.2 槽底放线

### 6.3.2 Setting-out at Trench Bottom

垫层打好后，用经纬仪或拉细线绳挂垂球，把龙门板或控制桩上的轴线投测到垫层上，如图 6-11 所示，用墨线弹出墙体中心线和基础边线，以便砌筑基础。整个墙体形状及大小均以该线为准。它是确定建筑物位置的关键环节，必须严格校核。

After the cushion is made, the axis on the batter board or control peg is transferred onto the cushion using a theodolite or a string with a plumb bob, as shown in Figure 6-11. The center line of the wall and the foundation boundary line are snapped with ink for foundation construction. The wall shape and dimensions shall be subject to the lines. Since it is a key link in determining the location of building, the lines must be strictly checked.

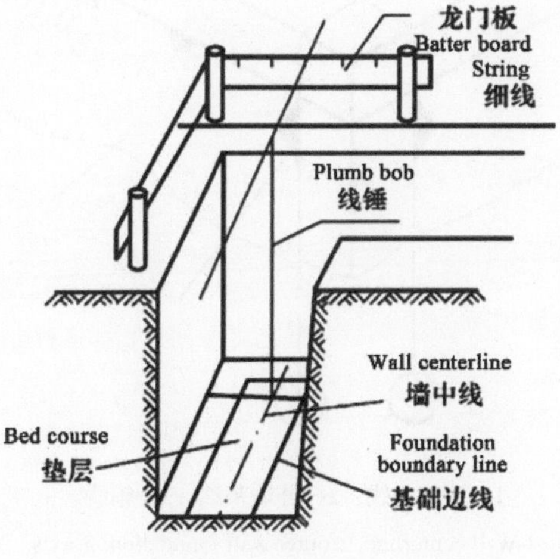


图6-11 槽底放线

Figure 6-11 Setting-out at Trench Bottom

## 6.3.3 基础墙砌筑时的标高控制

### 6.3.3 Elevation Control During Foundation Wall Masonry

用砖砌筑基础墙时，一般采用基础皮数杆作为标高控制依据。基础皮数杆是一根木制的杆子（图 6-12）。按照设计尺寸，在杆子上将砖厚度、灰缝厚度、层数画出，并标明 ±0.000、防潮层等的标高位置。

During the masonry of brick foundation wall the height pole is generally used as the basis for elevation control. It is a wood pole (Figure 6-12). According to the design size, the thickness of bricks, the thickness of mortar joints and the number of layers shall be drawn on the pole, and the elevations such as ±0.000 and moisture-proof layer elevations are marked.

在立皮数杆处打一大木桩，用水准仪在该木桩上测设一条比垫层顶标高高（如 10 cm）的水平线。将皮数杆上标高相同的一条线与木桩上的水平线对齐，并用大钉把皮数杆钉到木桩上，作为砌砖时的标高依据。

A large wood pile is driven at a place where the height pole is erected, and a horizontal line (e.g. 10 cm) higher than the top elevation of the cushion is measured and set on the wood pile with a level gauge. Align a line with the same elevation on the

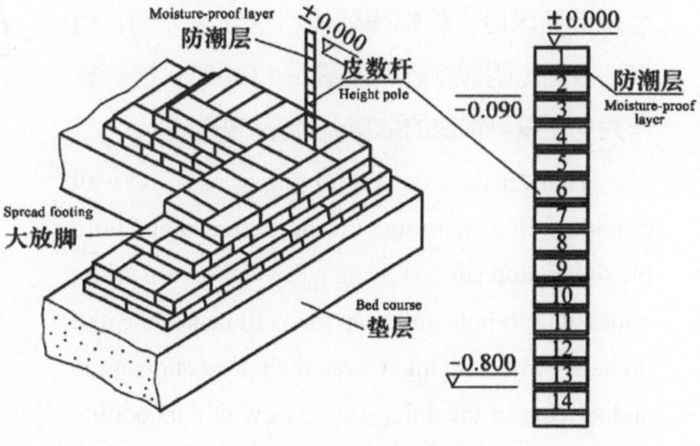


图6-12 基础皮数杆

Figure 6-12 Foundation Height Pole

height pole with the horizontal line on the wood pile, and spike the height pole into the wooden pile as the elevation basis during bricklaying.

用毛石砌筑基础墙时，在龙门板中心钉上挂细线，以控制毛石基础墙的高度。

During the masonry of rubble foundation wall, a thin line shall be hung on the center nail of the batter board to control the height of the masonry rubble foundation.

基础墙砌筑完后，用水准仪检查各轴线交点上的基础面的标高是否符合设计要求。一般民用建筑物的基础面标高容许误差范围不大于 ± 10 mm。

After the foundation wall is built, check with a level gauge the elevations of the foundation surfaces at the intersections of axes to meet the design requirements. The allowable error range of foundation surface elevation of general civil buildings is not more than ±10 mm.

## 6.4 墙体工程施工测量

## 6.4 Construction Survey of Wall Works

墙体施工中的测量工作包括墙体定位弹线及提供墙体各部位的标高。

The survey work during wall construction includes wall positioning, snapping lines, and determination of elevations of various parts of the wall.

用经纬仪或拉细线绳挂垂球，将经检查无误后的轴线控制桩或龙门板上的轴线和墙边线标志投测到基础面上，用墨线弹出墙体中线与边线。用经纬仪检查外墙轴线交角及间距，符合规范要求后，将轴线延伸到基础墙外侧，用红油漆画出明显标志，如图 6-13 所示。该标志作为向上投测轴线的依据，应保护好。

Transfer the axis control peg or axis and wall boundary line markings on the batter board onto the foundation surface using a theodolite or a string with a plumb bob, and snap the wall centerline and boundary line with ink. Check the intersection angle and spacing of the outer wall axes with a theodolite to meet the specification requirements, then, extend the axis to the outside of the foundation wall and make a clear mark with red paint, as shown in Figure 6-13. This mark, as a reference for upward axis transfer, must be protected properly.

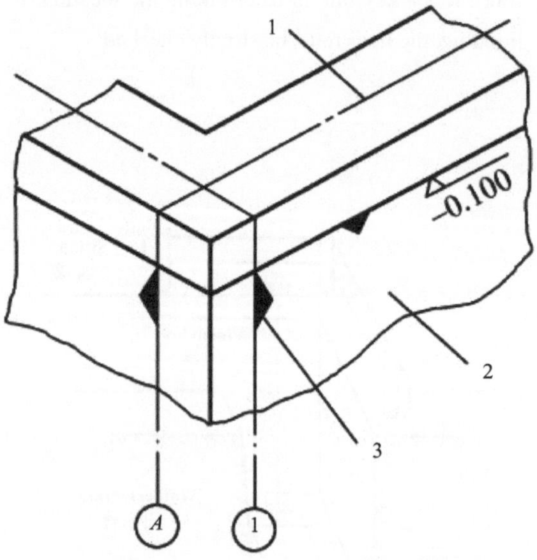


1—墙体中线；2—外墙基础；3—轴线。

1-wall centerline; 2-outer wall foundation; 3-axis.

图6-13　墙体轴线与标高线

Figure 6-13 Wall Axis and Elevation Line

另外，在基础墙外侧画出门、窗和其他预留洞的边线。

In addition, draw the boundary lines of doors, windows and other reserved holes on the outside of the foundation wall.

在用砖砌筑基础墙时，墙体各部位标高通常也采用皮数杆（图 6-14）来控制。在皮数杆上依据设计尺寸，按砖和灰缝厚度画线，并标明门、窗、过梁、楼板等的标高位置。杆上标高注记从 ±0.000 m 向上增加。墙身皮数杆一般立在建筑物的拐角和内墙处，若采用内脚手架施工，皮数杆应立在外侧；反之，皮数杆应立在内侧。立皮数杆时，先用水准仪在立杆处的木桩或基础墙上测设出 ±0.000 m 标高线，测量误差在 ±3 mm 以内，然后把皮数杆上的 ±0.000 m 线与该线对齐，用吊锤校正并用钉钉牢，必要时可在皮数杆上加两根钉斜撑，以保证皮数杆的稳定。

During the masonry of brick foundation wall, the elevation of each part of the wall is usually controlled by height pole (Figure 6-14). Draw lines on the height pole based on the design size and the thickness of bricks and mortar joints, and mark the elevation positions of doors, windows, lintels, floors, etc. The elevation on the pole is increased from ±0.000 m. The wall height pole is generally erected at the corner and inner wall of the building. If the inner scaffold is used for construction, the height pole shall be erected on the outside; otherwise, the height pole shall be erected on the inside. To erect the height pole, first measure ±0.000 m elevation line on the wooden pile or foundation wall at the pole position using a level gauge, and the measurement error is within ±3 mm. Then align the ±0.000 m line on the height pole with the elevation line, correct it with a drop-hammer and nail it firmly. If necessary, two diagonal bracings may be nailed to the height pole to ensure stability.

当墙体砌到窗台时，用水准仪在室内外墙体上测设一条 +0.5 m 的标高线并弹出墨线，称为“+50 线”。外墙的 +50 线可作为向上传递各楼层标高的依据，内墙的 +50 线可作为室内地面施工及室内装修的标高依据。

When the wall is built to the window sill, a +0.5 m elevation line is measured on the inner and outer walls with a level gauge and an ink line is snapped, which is referred to as "+50 line". The +50 line on the outer wall is used as the basis for transferring the elevation of each floor upward, and the +50 line on the inner wall is used as the elevation basis for indoor ground construction and indoor decoration.

墙的垂直度是用如图 6-15 所示的托线板来进行检查的。把托线板紧靠墙面，如果铅垂线与板上的墨线不重合，就要对砌砖的位置进行校正。

The perpendicularity of the wall is checked with a cord gauge as shown in Figure 6-15. Put the cord gauge against the wall closely. If the plumb line does not coincide with the ink line on the cord gauge, the position of the brickwork should be corrected.

在楼板安装好后，将底层墙体轴线引测到楼面上，并定出墙边线。用水准仪测出楼面四角标高，取平均值作为地坪标高。当精度要求较高时，用钢尺从 + 0.5 m 线向上直接丈量到该层楼板外侧，作为重新立皮数杆的标高标志。框架或钢筋混凝土柱、间墙施工时，在每层的柱、墙上测设 +0.5 m 标高线代替皮数杆，作为标高控制的依据。

After the floor slab is installed, the axis of the wall on the ground floor shall be transferred to the floor and the wall boundary line shall be set out. Measure the elevations at the four corners of the floor with a level, and take the average value as the floor elevation. When there is a high accuracy requirement, use a steel ruler to directly measure from the +0.5 m line to the outside of the floor slab as an elevation mark for re-erection of the height pole. During the construction of frames or reinforced concrete columns and partition walls, an +0.5 m elevation line shall be measured respectively on the columns and walls on each floor to replace the height pole, as the basis for elevation control.

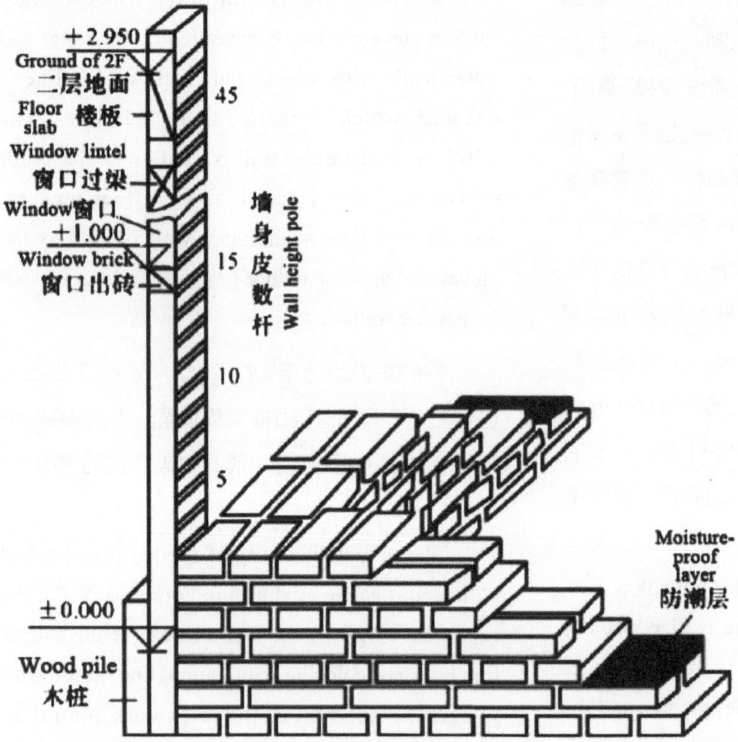


图6-14 墙身皮数杆

Figure 6-14 Wall Height Pole

图6-15 托线板

Figure 6-15 Cord Gauge

# 本章小结

## Chapter Summary

本章主要讲述了民用建筑施工测量前的准备工作、建筑物的定位与放线、建筑物基础工程施工测量等内容。

This chapter mainly describes the preparation before construction survey of civil buildings, positioning and setting-out of buildings, construction survey of building foundation works, etc.

1. 在进行民用建筑施工测量前，应做好的测设准备工作有熟悉图纸、仪器配备与检校、现场踏勘、编制施工测设方案和准备测设数据。

1.Before the construction survey of civil buildings, you should get familiar with the drawings, prepare and calibrate instruments, conduct site survey, and prepare construction survey schemes and data.

2. 民用建筑施工测量的工作内容主要有建筑物的定位与放线、基础工程施工测量、墙体工程施工测量、各层轴线投测及标高传递等。

2.The construction survey of civil buildings mainly includes the positioning and setting-out of buildings, the construction survey of foundation works, the construction survey of wall works, the axis and elevation transfer of each floor.

# 复习思考题

# Review Questions

## 一、填空题

## I. Fill in the Blanks

1. 施工测量时必须具备的图纸及资料有______________、______________、______________、______________、______________和______________。

1. The drawings and data that must be available during construction survey include______________, ______________, ______________,______________, ______________, and ______________.

2. 建筑物的放线就是______________。

2. The setting-out of building is______________.

3. 房屋基础墙（±0.000 以下的砖墙）的高度是利用______________来控制的。

3. The height of building foundation wall (brick wall below ±0.000) is controlled by______________.

## 二、选择题

## II. Choice Questions

1. 建筑平面图给出的是建筑物（ ）等。

1. The building plan shows ( ).

A. 各定位轴线间的尺寸关系

A. dimensional relationship between positioning axes

B. 基础轴线间的尺寸关系和编号

B. dimensional relationship and number between foundation axes

C. 室内地坪标高

C. indoor floor elevation

D. 基础、地坪、门窗、楼板、屋架和屋面等的设计高程

D. design elevations of foundation, floor, door and window, floor slab, roof truss and roof

2. 龙门板延长轴线法适用于一般的( )民用建筑物。

2.The axis extension using a batter board is suitable for ( ) civil buildings.

A. 小型 B. 中型 C. 大型 D. 所有

A. small B. medium C. large D. all

3. 采用里脚手架砌砖时，皮数杆应立在墙的( )；采用外脚手架时，皮数杆应立在墙的( )。

3. When the internal scaffold is used for brickwork, the height pole shall be erected on ( ) of the wall. When the external scaffold is used for brickwork, the height pole shall be erected on ( ) of the wall.

A. 内侧 内侧 B. 内侧 外侧 C. 外侧 内侧 D. 外侧 外侧

A. inside; inside B. inside; outside C. outside; inside D. outside; outside

## 三、简答题

## III. Short Answer Questions

1. 民用建筑施工测量前的准备工作有哪些？

1. What are the preparations before construction survey of civil buildings?

2. 设置龙门板或引桩的作用有哪些？

2. What are the purposes of a batter board or a guide pile?

3. 在建筑物基槽施工中，如何控制开挖深度？

3. How to control the excavation depth during the construction of building foundation trench?

7

建筑工程测量
ARCHITECTURAL ENGINEERING SURVEYING

# 建筑变形观测与竣工测量

# Building Deformation Observation and Acceptance Survey

**学习目标：**

**Learning Objectives:**

通过本章的学习，了解建筑变形观测的任务及内容；熟悉建筑物沉降观测水准点和观测点的布置要求、建筑工程竣工测量的内容；掌握建筑物沉降观测及观测结果整理的方法，建筑物倾斜观测、裂缝观测、水平位移观测方法，竣工总平面图编绘的步骤。

Through the study of this chapter, to learn about the tasks and contents of building deformation observation; to get familiar with the layout requirements of building settlement observation benchmarks and observation points, and the contents of construction project acceptance survey; to master the methods and results sorting of building settlement observation, the methods of building inclination observation, crack observation and building horizontal displacement observation, and steps of compilation of general as-built plan.

**能力目标：**

**Ability Objectives:**

能正确使用仪器进行沉降观测、倾斜观测、裂缝观测、位移观测、特殊变形观测，能正确绘制建筑物竣工总平面图。

Be able to correctly use instruments for settlement observation, inclination observation, crack observation, displacement observation and special deformation observation, and draw the general as-built plan of the building.

# 7.1 建筑变形观测概述

## 7.1 Overview of Building Deformation Observation

民用建筑、工业厂房的施工过程中或使用期间，由于建筑物基础的地质构造不均匀，土壤的物理性质不同，受大气温度变化、土基的弹性变形、地下水位季节性和周期性的变化、建筑物本身的荷重及动荷载（如风力、震动）等因素的影响，建筑物将发生沉降、位移、倾斜及产生裂缝等变形现象。这些变形在一定范围内不会影响建筑物的正常使用，可视为正常现象，但如果超过一定限度就会影响建筑物的正常使用，严重的还会危及建筑物的安全。为使建筑物能被正常安全地使用，在建筑物施工各阶段及使用期间，应对建筑物进行有针对性的变形观测，通过变形观测可以分析和监视建筑物的变形情况。当发现异常变形时，应及时分析原因，采取相应补救措施，以确保施工质量，保证建筑物安全适用，同时，这也可为将来的设计与施工积累资料。

During the construction or use of civil buildings and industrial plants, due to the uneven geological structure of the building foundation, different physical properties of the soil, changes in atmospheric temperature, elastic deformation of the soil foundation, seasonal and periodic changes in groundwater level, and the load and dynamic load (such as wind force and vibration) of the building itself, settlement, displacement, inclination, cracks and other deformations will occur in the building. These deformations will not affect the normal use of the building within a certain limit and can be regarded as normal phenomena. However, if they exceed a certain limit, the normal use of the building will be affected, and even the building safety will be threatened in serious cases. In order to ensure the normal and safe use of the building, it is necessary to carry out targeted deformation observation at all construction stages and during use, and the building deformation can be analyzed and monitored through deformation observation. When abnormal deformation is found, the causes shall be analyzed in time and remedial measures shall be taken correspondingly to ensure the construction quality and safety, and at the same time, this can also accumulate data for future design and construction.

### 7.1.1 建筑变形观测的任务及内容

### 7.1.1 Tasks and Contents of Building Deformation Observation

建筑变形观测的任务是周期性地对建筑物上的观测点进行重复观测，以求得观测点位置的变化量。建筑变形观测的内容主要有沉降观测、位移观测、倾斜观测、裂缝观测等。

The task of building deformation observation is to repeatedly observe the observation points on the building periodically in order to obtain the variation quantity of the observation points. The building deformation observations mainly refer to settlement observation, displacement observation, inclination observation, crack observation, etc.

变形观测需要在不同时期多次进行，从历次观测结果的比较中，了解变形量随时间而变化的情况，它是监视工程建筑物在各种应力作用下是否安全的重要手段，其结果也是验证设计理论和检验施工质量的重要资料。

Deformation observation needs to be carried out many times in different periods. From the comparison of previous observation results, understanding the change of deformation with time is an important means to monitor the building safety under various stress effects, and its results are also important data to verify the design theory and to check the construction quality.

## 7.1.2 建筑变形观测的等级划分及精度要求

## 7.1.2 Classification and Accuracy Requirements of Building Deformation Observation

建筑变形观测的精度要求取决于该建筑物预计的允许变形值的大小和进行观测的目的。建筑变形观测的等级、精度指标及其适用范围，应符合表 7-1 的规定。

The accuracy of building deformation observation depends on the expected allowable deformation value of the building and the purpose of observation. The classification, accuracy index and applicable scope of building deformation observation shall comply with Table 7-1.

表 7-1 建筑变形观测的等级、精度指标及其适用范围

Table 7-1 Classification, Accuracy Index and Applicable Scope of Building Deformation Observation

| 等级<br>Class | 沉降监测点测站高差中误差/mm<br>Mean square error of elevation difference on settlement monitoring point station/mm | 位移监测点坐标中误差/mm<br>Mean square error of displacement monitoring point/mm | 主要适用范围<br>Main scope of application |
|---|---|---|---|
| 特等<br>Special | 0. 05 | 0. 3 | 特高精度要求的变形测量<br>Deformation measurement with extra-high accuracy requirement |
| 一等<br>First-order | 0. 15 | 1. 0 | 地基基础设计为甲级的建筑的变形测量，重要的古建筑、历史建筑的变形测量，重要的城市基础设施的变形测量等<br>Deformation measurement of buildings with foundation design of Class A, deformation measurement of important ancient buildings and historical buildings, deformation measurement of important urban infrastructure, etc |
| 二等<br>Second-order | 0. 5 | 3. 0 | 地基基础设计为甲、乙级的建筑的变形测量，重要场地的边坡监测，重要的基坑监测，重要管线的变形测量，地下工程施工及运营中的变形测量，重要的城市基础设施的变形测量等<br>Deformation measurement of buildings with foundation design of Class A and Class B, slope monitoring of important sites, important foundation pit monitoring, deformation measurement of important pipelines, deformation measurement during underground construction and operation, deformation measurement of important urban infrastructure, etc |

续表

| 等级<br>Class | 沉降监测点测站高差中误差/mm<br>Mean square error of elevation difference on settlement monitoring point station/mm | 位移监测点坐标中误差/mm<br>Mean square error of displacement monitoring point/mm | 主要适用范围<br>Main scope of application |
|---|---|---|---|
| 三等<br>Third-order | 1. 5 | 10 | 地基基础设计为乙、丙级的建筑的变形测量，地表、道路及一般管线的变形测量，一般的城市基础设施的变形测量，日照变形测量，风振变形测量等<br>Deformation measurement of buildings with foundation design of Class B and Class C, deformation measurement of ground surface, roads and general pipelines, deformation measurement of general urban infrastructure, sunshine deformation survey, wind vibration deformation measurement, etc. |
| 四等<br>Fourth-order | 3 | 20 | 精度要求低的变形测量<br>Deformation measurement with low accuracy requirement |

注：1.沉降监测点测站高差中误差：对于水准测量，为其测站高差中误差；对于静力水准测量、三角高程测量，为相邻沉降监测点间等价的高差中误差；

Note: 1. Mean square error of elevation difference on settlement monitoring point station: for leveling, it is the mean square error of elevation difference on the observation station; for hydrostatic leveling and trigonometric leveling, it is the equivalent mean square error of elevation difference between adjacent settlement monitoring points;

2.位移监测点坐标中误差：其是指监测点相对于基准点或工作基点的坐标中误差、监测点相对于基准线的偏差中误差、建筑上某点相对于其底部对应点的水平位移分量中误差等。

2. Mean square error of displacement monitoring point coordinate: It refers to the mean square error of coordinate of the monitoring point relative to the datum point or the working reference point, the mean square error of deviation of the monitoring point relative to the datum line, and the mean square error of horizontal displacement component of a certain point on the building relative to the corresponding point at its bottom.

# 7.2 建筑物的沉降观测

## 7.2 Settlement Observation of Building

在建筑物施工过程中，随着上部结构的逐步建成，地基荷载的逐步增加，建筑物将出现下沉现象。为了掌握建筑物的沉降情况，及时发现对建筑物不利的下沉现象，在建筑物施工过程中和建筑物投入使用后，必须对建筑物进行沉降观测。

During building construction, as the superstructure is completed gradually, the foundation load will increase accordingly, which will cause uneven subsidence of building. In order to detect the uneven and unfavorable settlement of the building in a timely manner, the building must be subject to settlement observation during the construction and use thereof.

### 7.2.1 水准基点和观测点的布设

### 7.2.1 Layout of Benchmarks and Observation Points

#### （1）水准基点的布设

#### (1) Layout of benchmarks

水准基点是固定不动的，可作为沉降观测高程的基准点。水准基点作为沉降观测的依据，必须保证其高程在相当长的观测时期内固定不变。其布设除应满足一般要求（布设在坚实稳固之处，底部应埋设在冻土层以下）外，还应满足以下特殊要求。

The benchmark is fixed and used as the elevation benchmark for settlement observation. As the basis for settlement observation, benchmark elevation must be fixed for a long observation period. The layout shall meet the general requirements (it shall be laid in a solid and stable place, and the bottom shall be buried below the frozen soil layer) and special requirements as follows.

①为了相互校核，保证沉降观测中所使用的水准基点的可靠性，防止由于水准基点的高程变化造成差错，对于特等、一等沉降观测，基准点不应少于四个，对于其他等级沉降观测，基准点不应少于三个，且基准点之间应形成闭合环。

① In order to check each other, ensure the reliability of the benchmarks used in settlement observation and prevent errors caused by elevation changes of the benchmarks, there shall be no less than 4 benchmarks for special and first order settlement observation, and no less than 3 benchmarks for settlement observation of other classifications, and a closed loop shall be formed by the benchmarks.

②水准基点应埋设在施工建筑的应力影响范围之外，不受打桩、机械加工和开挖等操作的影响。

② The benchmarks shall be buried beyond the influence range of stress of the construction building and shall not be affected by piling, machining and excavation.

③应根据土质情况和在了解建筑物预计沉降量的情况下，选择水准基点的形式。对于稳定的原状土层，一般可选用混凝土普通标石，也可以将标志镶嵌在裸露的基岩上。若受条件限制，在变形区内也可埋设深层钢管标志等。

③ The form of benchmark shall be selected according to the soil conditions and the expected settlement of the building. For stable undisturbed soil layer, ordinary concrete markstone can be generally chosen, or marks can be embedded on exposed bedrock. If there is a condition limit, steel pipe marks may be buried deeply in the deformation area.

#### （2）观测点的布设

#### (2) Layout of observation points

观测点是为进行沉降观测而设置在建筑物上的固定标志。建筑物的沉降量是通过水准测量方法测定的，即通过多次观测水准基点与设置在建筑物上的观测点之间的高差的变化测定建筑物的沉降量。为了能全面、准确地反映整个建筑物的沉降变化情况，必须合理确定观测点的数目和位置。

Observation points are fixed markers arranged on the building for settlement observation. The building settlement is determined by leveling, that is, the settlement of building is determined by observing the change in the elevation difference between the benchmark and the observation point on the building for many times. In order to reflect the variation of settlement of the whole building comprehensively and accurately, the number and location of observation points must be reasonably determined.

在民用建筑中，通常是在房屋的转角、沉降缝两侧、基础变化处，以及地质条件改变处设置观测点。一般在建筑物的四周每隔 10~20 m 设置一沉降观测点。

In civil buildings, observation points are usually set at the corners of buildings, on both sides of settlement joints, and where the foundation changes and the geological conditions change. Generally, settlement observation points are set every 10-20 m around the building.

建筑物的宽度较大时，还应在房屋内部纵墙上或楼梯间布置观测点。

When the width of the building is large, observation points shall also be arranged on the longitudinal wall or staircase inside the building.

工业厂房可在柱子、承重墙、厂房转角、大型设备基础的周围设置观测点。扩建的厂房应在连接处两侧基础墙上设置观测点。

For industrial plants, observation points can be set around columns, bearing walls, plant corners and large equipment foundations. Observation points shall be set on the foundation walls on both sides of the connection of the expanded plant.

对于高大的圆形构筑物（水塔、高炉、烟囱等），应在其基础的对称轴上设置观测点。

For tall circular structures (water tower, blast furnace, chimney, etc.), observation points shall be arranged on the symmetry axes of their foundations.

观测点的高度、方向要便于立尺和观测，不宜受到破坏。至于承重墙和柱子上的观测点的埋设形式，一般民用建筑的沉降观测点设置方法是用长 120 mm 左右的角钢，一端焊接一铆头，另一端埋入墙或柱内，用 1 ：2 水泥砂浆填实，如图 7-1(a) 所示；或者用钢筋，一端弯成向上的直钩埋入墙或柱内，如图 7-1(b) 所示；基础上的观测点通常是将铆钉或钢筋埋入混凝土中，如图 7-2 所示。

The height and direction of observation points shall make it easy to set up the steel ruler and observe, and shall not be damaged. As for the arrangement of observation points on bearing walls and columns, the settlement observation points of general civil buildings are arranged as follows: one end is welded with a rivet head using 120 mm long angle steel, and the other end is embedded in the wall or column, and filled with 1 ：2 cement mortar, as shown in Figure 7-1(a); alternatively, one end of reinforcement bars is bent into an upward

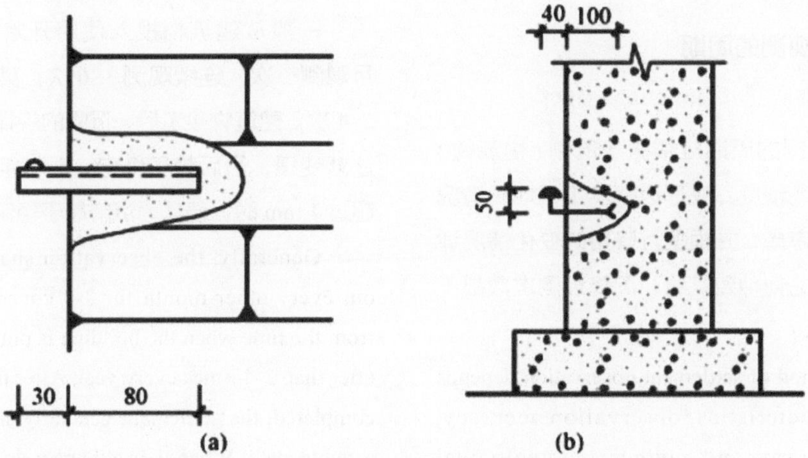


图7-1　墙体沉降观测点的埋设

Figure 7-1 Embedment of Wall Settlement Observation Points

straight hook and buried in the wall or column, as shown in Figure 7-1(b); for observation points on the foundation, rivets or reinforcement are usually embedded in concrete, as shown in Figure 7-2.

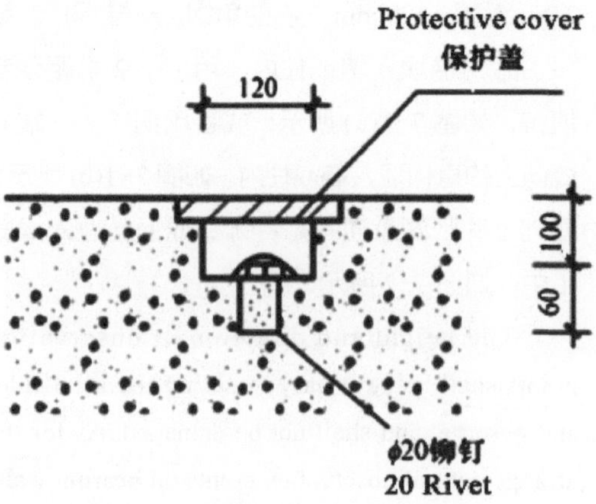


图7-2　设备基础沉降观测点的埋设

Figure 7-2 Embedment of Equipment Foundation Settlement Observation Points

## 7.2.2 沉降观测的周期、方法和精度

## 7.2.2 Period, Method and Accuracy of Settlement Observation

### （1）沉降观测的周期

### (1) Period of settlement observation

沉降观测的周期应根据建筑物（构筑物）的特征、观测精度、变形速率及工程地质情况等综合因素考虑，并根据沉降量的变化情况适当调整。无论何种建筑物，沉降观测次数都不能少于五次。

The period of settlement observation depends on the characteristics, observation accuracy, deformation rate and engineering geological conditions of buildings (structures), and shall be adjusted appropriately based on the variation of settlement. No matter what kind of building, the settlement observation shall be done for no less than five times.

建筑物施工期间，当观测点安置稳定后，应及时进行第一次沉降观测。在每次增加较大荷载前后，如基础回填土、墙体每砌筑一层楼、屋架安装、设备安装与运行、塔体加高一层、烟囱每加高 15 m 时都应进行沉降观测。在施工过程中若停工时间较长，应在停工时和复工前各进行一次观测。竣工后应根据沉降量的大小来确定观测的时间间隔。

During building construction, the first settlement observation shall be carried out timely after the observation points are placed stably. Settlement observation shall be done before and after each increase of large load, such as foundation backfilling, wall masonry on each floor, roof truss installation, equipment installation and operation, tower height increase by one floor, and chimney height increase by every 15 m. If the work is suspended for a long time, observation shall be made at the time of suspension and before the resumption of work. The observation interval shall be determined based on the settlement amount after completion.

一般从建筑物投入使用开始，每隔一个月观测一次，连续观测 3~6 次，其后一年观测 2~4 次。建筑物竣工后，沉降的规律一般是由快逐渐变缓，最后趋于稳定。当半年内沉降量不超过 1 mm 时，便认为沉降趋于稳定。

Generally, the observation shall be carried out every other month for 3-6 consecutive times from the time when the building is put into use, and after that, 2-4 times every year. After the building is completed, the settlement generally slows down to a stable state. When the settlement does not exceed 1 mm within half a year, it is considered to be stable.

## （2）沉降观测的方法

### (2) Settlement observation method

沉降观测常采用水准测量的方法，根据布设好的水准基点和沉降观测点，布设闭合或者附合水准路线，采用水准路线计算方法对高差闭合差进行分配，从而对计算结果进行检校，最终计算出水准路线上的每个观测点的高程。

Leveling is often adopted for settlement observation. According to the benchmarks and settlement observation points, closed or annexed leveling lines are arranged, and the elevation difference closure error is distributed using the leveling line calculation method, so as to check the calculation results and finally calculate the elevation of each observation point on the leveling line.

一般来说，对于高层建筑物的沉降观测应使用 $DS_1$ 精密水准仪，按国家二等水准测量方法和精度进行；对于多层建筑物的沉降观测应使用 $DS_3$ 水准仪，按普通水准测量的方法和精度进行。

Generally, $DS_1$ precision level shall be used for settlement observation of high-rise buildings according to the national second-order leveling method and accuracy; $DS_3$ level shall be used for settlement observation of multi-story buildings according to the general leveling method and accuracy.

观测过程中应重视第一次观测的结果，因为首次观测的高程是以后各次观测用以比较的依据；若初测精度低，会造成后续观测数据存在误差。为保证初测精度，应进行两次观测，每次均布设成闭合水准路线，以闭合差来评定观测精度。

Attention shall be paid to the results of the first observation during the observation, because the elevation of the first observation is the basis for comparison of subsequent observations. If the accuracy of the first observation is low, it will cause errors in the subsequent observation data. In order to ensure the accuracy of the first observation, two observations shall be done, and for each time, a closed leveling line shall be formed to evaluate the observation accuracy based on the closure error.

## （3）沉降观测的精度

### (3) Accuracy of settlement observation

为保证沉降观测的精度，减小仪器工具、设站等方面的误差，一般采用同一台仪器、同一根标尺，每次在固定位置架设仪器，固定观测几个观测点和固定转点位置，同时应注意使前、后视距相等，以减小角误差的影响。

In order to ensure the accuracy of settlement observation and reduce the errors of instruments and observation stations, the same instrument and the same ruler are generally used; the instrument is set up at a fixed position each time; several observation points and turning points are fixed. At the same time, the foresight distance and backsight distance should be equal to reduce the influence of angle error.

沉降观测时，应从水准基点开始，组成闭合或附合路线逐点观测。对于连续生产设备基础和动力设备基础、高层钢筋混凝土框架及不均匀地基上的重要建筑物，沉降观测的水准闭合差不应超过 $\pm\sqrt{N}$ mm（$N$ 为测站数），同一后视点两次后视读数之差不应超过 ±1 mm。对于一般多层建筑物、厂房基础和构筑物的沉降观测，往返观测水准点的高差较差不应超过 $\pm 2/\sqrt{N}$ mm（$N$ 为测站数），同一后视点两次后视读数之差不应超过 ±2 mm。

During settlement observation, a closed or annexed leveling line shall be formed from the benchmark for observation point by point. For continuous production equipment foundation, power equipment foundation, high-rise reinforced concrete frame and important buildings on uneven foundation, the error of closure in leveling for settlement observation shall not exceed ± $\sqrt{N}$ mm ($N$ is the number of observation stations), and the difference between two backsight readings at the same backsight point shall not exceed ±1mm. For settlement observation of general multi-storey buildings, plant foundations and structures, the elevation difference between direct and reversed observation benchmarks shall not exceed ± $2/\sqrt{N}$ mm ($N$ is the number of observation stations), and the difference between two backsight readings at the same backsight point shall not exceed ± 2 mm.

## 7.2.3 沉降观测的注意事项

## 7.2.3 Precautions for Settlement Observation

为了保证沉降观测满足上述的精度要求，必须注意以下几点。

In order to ensure that the settlement observation meets the above accuracy requirements, attention shall be paid to the following.

①施测前，应对测量仪器进行严格的检查校正。

① Before measurement, the measuring instruments must be strictly checked and calibrated.

②应尽可能在不转站的情况下测出各观测点的高程，前后视距应尽量相等，整个观测最好用同一根水准尺，观测应在成像清晰、稳定的条件下进行，避免阳光直射仪器。

② The elevation of each observation point should be measured at the same point as far as possible, and the foresight distance and backsight distance should be as equal as possible. The same leveling rod should preferably be used throughout the observation, and the observation should be carried out when the imaging is clear and stable. The instruments shall be protected against direct sunlight.

③测量中，应尽量做到观测人员固定、测量仪器固定、水准基点固定、测量路线固定及测量方法固定。

③ The observers, measuring instruments, benchmarks, leveling lines and leveling methods should be fixed as far as possible.

## 7.2.4 沉降观测的结果整理

## 7.2.4 Results Sorting of Settlement Observation

每一观测周期结束后，应及时对观测结果进行整理，计算观测点的沉降量、沉降差以及本周期平均沉降量和沉降速度。

After each period of observation, the observation results shall be sorted out in time to calculate the settlement amount and settlement difference of the observation points, and the average settlement amount and settlement speed in this period.

### （1）整理原始记录

### (1) Sorting out original records

每次观测结束后，应检查记录的数据和计算是否正确，精度是否合格，然后调整高差闭合差，推算出各沉降观测点的高程，并填入沉降观测记录表中（表 7-2）。

After each observation, it is necessary to check whether the recorded data and calculations are correct and the accuracy meets the requirements, and then adjust the closure error of elevation difference, calculate the elevation of each settlement observation point, and fill in the Settlement Observation Record Form (Table 7-2).

表 7–2　沉降观测记录表

Table 7-2 Settlement Observation Record Form

| 观测次数 Observation frequency | 观测时间 Observation time | 各观测点的沉降情况 Settlement of each observation point | | | | | | | 施工进展情况 Construction-progress | 荷载情况 / (t · m$^{-2}$) Loads/ (t · m$^{-2}$) |
|---|---|---|---|---|---|---|---|---|---|---|
| | | 1 | | | 2 | | | … | | |
| | | 高程/m Elevation /m | 本次下沉 / mm Settlement this time / mm | 累积下沉 / mm Accumulated settlement /mm | 高程/m Elevation /m | 本次下沉 / mm Settlement this time / mm | 累积下沉 / mm Accumulated settlement/ mm | … | | |
| 1 | 2001-01-10 | 50. 454 | 0 | 0 | 50.473 | 0 | 0 | | 一层平口 1F completed | |
| 2 | 2001-02-23 | 50. 448 | −6 | −6 | 50.467 | −6 | −6 | | 三层平口 3F completed | 40 |
| 3 | 2001-03-16 | 50. 443 | −5 | −11 | 50. 462 | −5 | −11 | | 五层平口 5F completed | 60 |
| 4 | 2001-04-14 | 50. 440 | −3 | −14 | 50.459 | −3 | −14 | | 七层平口 7F completed | 70 |
| 5 | 2001-05-14 | 50. 438 | −2 | −16 | 50.456 | −3 | −17 | | 九层平口 9F completed | 80 |
| 6 | 2001-06-04 | 50. 434 | −4 | −20 | 50.452 | −4 | −21 | | 主体完工 Main works completed | 110 |
| 7 | 2001-08-30 | 50. 429 | −5 | −25 | 50. 447 | −5 | −26 | | 竣工 Completion | |
| 8 | 2001-11-06 | 50. 425 | −4 | −29 | 50.445 | −2 | −28 | | 使用 Use | |
| 9 | 2002-02-28 | 50. 423 | −2 | −31 | 50. 444 | −1 | −29 | | | |
| 10 | 2002-05-06 | 50. 422 | −1 | −32 | 50. 443 | −1 | −30 | | | |
| 11 | 2002-08-05 | 50. 421 | −1 | −33 | 50.443 | 0 | −30 | | | |
| 12 | 2002-12-25 | 50. 421 | 0 | −33 | 50.443 | 0 | −30 | | | |

注：水准点的高程，$BM_1$为 49. 538 mm, $BM_2$ 为 50. 123 mm, $BM_3$ 为 49. 776 mm。

Note: Elevations of benchmarks, $BM_1$ is 49. 538 mm, $BM_2$ is 50. 123 mm, $BM_3$ is 49. 776 mm.

### （2）计算沉降量

### (2) Calculation of settlement

计算各观测点本期沉降量（各观测点本次观测所得高程减去其上次观测所得高程）、累积沉降量（各观测点本期沉降量加上其上次累积沉降量）和观测日期、施工进度、荷载等情况，填入表 7–2 中。

Calculate the settlement of each observation point in the current period (the elevation this time minus the last elevation), the cumulative settlement (the settlement of all observation points in the current period plus the cumulative settlement of the previous observation), and the observation date, construction progress, load, etc., and fill in Table 7-2.

### （3）绘制沉降曲线

### (3) Drawing of the settlement curve

当变形观测进行到一定周期，或是工程进度到一定阶段，就要依据前面所监测和计算的结果，绘制点位时间与荷载、沉降量关系曲线图，如图 7–3 所示。通过变形曲线可以直观地了解变形过程情况，也可以对变形发展趋势有个直观的判断。

When the deformation observation reaches a certain period or the project progress reaches a certain stage, the curve chart of point time, load and settlement shall be drawn according to the results of the above observation and calculation, as shown in Figure 7-3. We can see intuitively the deformation process from the curve and have an intuitive judgment on the deformation development trend.

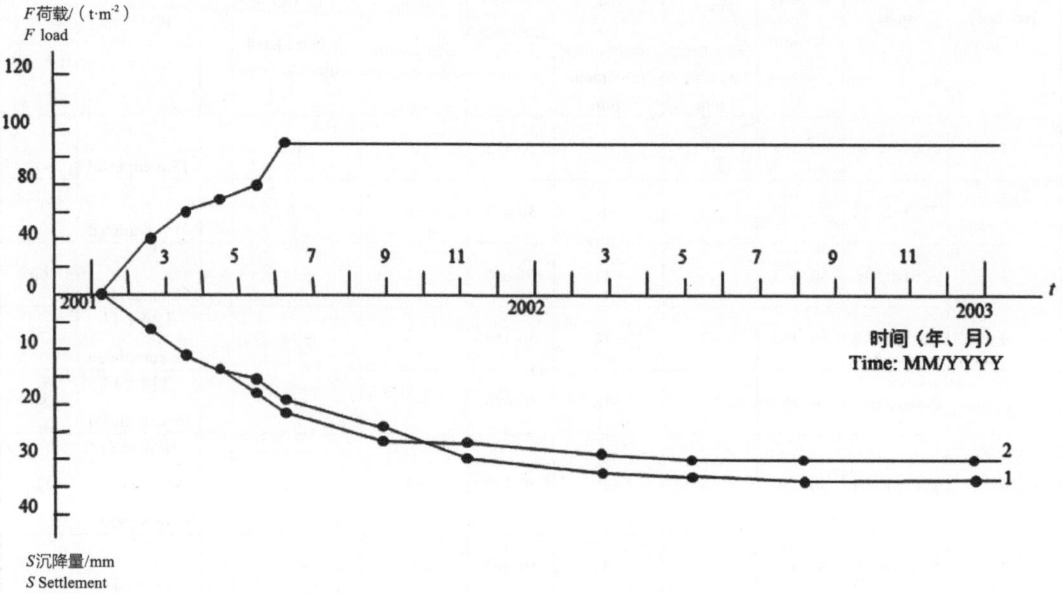


图7-3 荷载、时间、沉降量关系曲线示意图

Figure 7-3 Relation Curve of Load, Time and Settlement

# 7.3 建筑物的倾斜观测

## 7.3 Inclination Observation of Building

建筑物受施工中的偏差以及不均匀沉降等因素的影响，会产生倾斜；对建筑物倾斜程度进行测量的工作即倾斜观测。

The building will be inclined due to deviation and differential settlement in construction; inclination observation means to measure the inclination of the building.

### 7.3.1 建筑物的倾斜观测

### 7.3.1 Inclination Observation of Buildings

建筑物产生倾斜的原因主要是地基承载力的不均匀、建筑物结构复杂形成不同荷载及受外力风荷、地震等影响引起建筑物基础的不均匀沉降。测定建筑物倾斜度随时间而变化的工作叫作倾斜观测。倾斜观测一般是用水准仪、经纬仪、垂球或其他专用仪器来测量建筑物的倾斜度。

The main reasons for building inclination include uneven bearing capacity of foundation, complex structure of building, and uneven settlement of building foundation caused by external wind load and earthquake. The inclination observation involves measuring the inclination of building with time. Generally, we measure the building inclination with a level, theodolite, plumb or other special instruments.

## （1）倾斜观测的方法

### (1)Method of inclination observation

根据倾斜观测原理，利用仪器测量出建筑物顶部或底部的倾斜位移值$\Delta s$，再计算出建筑物的倾斜度，即

According to the inclination observation principle, the inclination displacement value $\Delta s$ of the top or bottom of the building is measured by the instrument, and then the inclination of the building is calculated, i.e.

$$i = \frac{\Delta s}{H} \tag{7.1}$$

式中，$i$为建筑物倾斜度；$H$为建筑物高度。

Where, $i$ is the inclination of the building; $H$ is the height of the building.

由式（7.1）可知倾斜测量主要是测定建筑物主体的倾斜位移值$\Delta s$，具体有如下几种方法。

It can be seen from Equation (7.1) that the inclination measurement is mainly to measure the offset value $\Delta s$ of the main body of the building, and there are several methods as follows.

①经纬仪观测法。

① Theodolite observation method.

利用经纬仪在两个互相垂直的方向上进行交会投点，将建筑物向外倾斜的一个上部角点投影至平地，直接量取其与下部角点的倾斜位移值分量$\Delta A$、$\Delta B$，则倾斜位移值$\Delta s = \sqrt{\Delta A^2 + \Delta B^2}$，如图 7-4 所示。

Use a theodolite to project an upper angular point of the building tilting outward to the flat ground in two perpendicular directions, and directly measure the tilt displacement components $\Delta A$ and $\Delta B$ with the lower angular point, so that the tilt displacement value $\Delta s = \sqrt{\Delta A^2 + \Delta B^2}$, as shown in Figure 7-4.

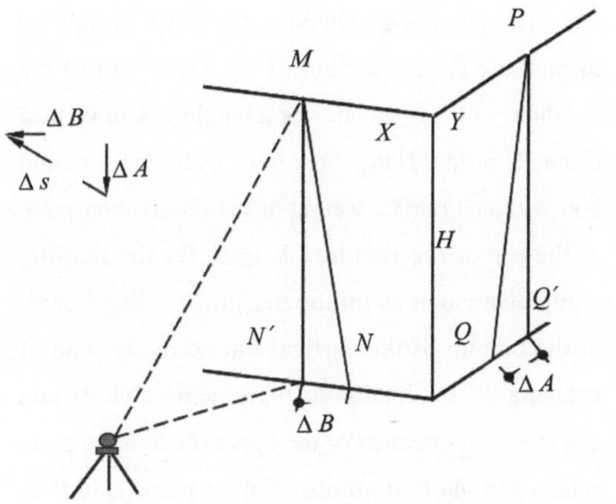


图7-4　经纬仪投点法倾斜观测

Figure 7-4 Inclination Observation with Theodolite Projection Method

②铅垂观测法。

② Vertical observation method.

当利用建筑物或构件的顶部与底部之间一定竖向通视条件进行观测时，宜选用铅垂观测方法。其又分为吊垂球法，正、倒垂线法，激光铅直仪观测法，激光位移计自动测记法等。

When the vertical intervisibility condition between the top and bottom of the building or component are available, the vertical observation method should be selected, including plumbing method, direct and inverse plummet observation method, laser plummet observation method, laser displacement meter automatic measurement method, etc.

a. 吊垂球法。

a. Plumbing method.

吊垂球法是测量建筑物上部倾斜的最简单方法，适合于内部有垂直通道的建筑物。在顶部或需要高度处的观测点位置上，直接或支出一点悬挂适当重量的垂球，在垂线下的底部固定读数设备（如毫米格网读数板），直接读取或量出上部观测点相对底部观测点的倾斜位移值 $\Delta s$ 和位移方向，再计算倾斜度 $i$。

The plumbing method is the simplest method to measure the inclination of the upper part of the building, which is suitable for a building with vertical channels inside. Hang directly or overhang a plumb bob with appropriate weight at the observation point at the top or the required height, fix the reading equipment (such as millimeter grid reading board) at the bottom of the vertical line, directly read or measure the inclination displacement value $\Delta s$ and displacement direction of the upper observation point relative to the bottom observation point, and then calculate the inclination $i$.

b. 正、倒垂线法。

b. Direct and inverse plummet observation method.

垂线宜选用直径 0.6 ～ 1.2 mm 的不锈钢丝或因瓦合金丝，并采用无缝钢管保护。采用正垂线法时，垂线上端可锚固在通道顶部或在所需高度处设置的支点上。采用倒垂线法时，垂线下端可固定在锚块上，上端设浮筒。用来稳定重锤和浮子的油箱中应装有阻尼液。观测时，由观测墩上安置的坐标仪、光学垂线仪、电感式垂线仪等量测设备，按一定周期测出各测点的水平位移量。

The plumb line should be stainless steel wire or invar alloy wire of 0.6-1.2 mm diameter and protected by seamless steel pipe. When the direct plumb line method is adopted, the upper end of the plumb line can be anchored at the top of the channel or at the fulcrum set at the required height. When the inverse plummet observation method is adopted, the lower end of the plumb line can be fixed on the anchor block, and the upper end is provided with a float. The tank used to stabilize the bob and float shall be filled with damping fluid. During observation, the horizontal displacement of each observation point is periodically measured by measuring equipment such as coordinate meter, optical plumb line meter and inductive plumb line meter installed on the observation pillar.

c. 激光铅直仪观测法。

c. Laser plummet observation method.

在顶部适当位置安置接收靶，在其垂线下的地面或地板上安置激光铅直仪或激光经纬仪，按一定周期观测，在接收靶上直接读取或量出顶部的水平位移量和位移方向，计算倾斜度 $i$。作业中仪器应严格置平、对中，应旋转 180° 观测两次取其中数。当仪器设在楼体内部时，应考虑大气湍流影响，如图 7-5 所示。

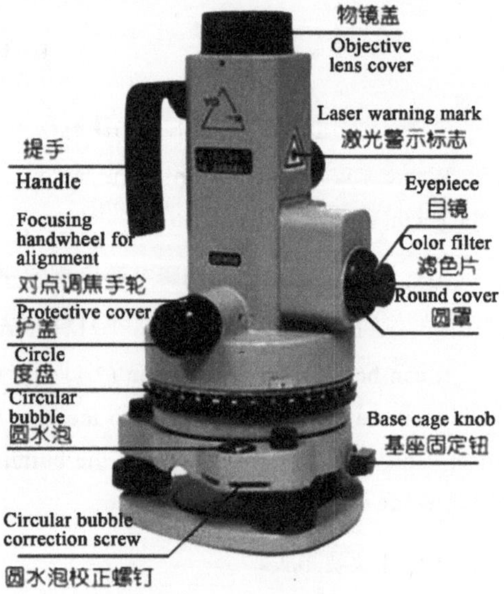


图7-5 激光铅直仪

Figure 7-5 Laser Plummet

Place the receiving target at an appropriate position on the top, place a laser plummet or laser theodolite on the ground or floor below the plumb line, observe periodically, directly read or measure the horizontal displacement and displacement direction of the top on the receiving target, and calculate the inclination $i$. During operation, the instrument shall be placed flat and centered, and shall be rotated by 180° to observe twice and take the mean. When the instrument is installed inside the building, the influence of atmospheric turbulence shall be considered, as shown in Figure 7-5.

d. 激光位移计自动测记法。

d. Automatic observation method with laser displacement meter.

用该法观测时，位移计安置在建筑物底层或地下室中，接收装置可设在顶层或需要观测的楼层，激光通道可利用楼梯间梯井，测试室宜选在靠近顶部的楼层内。当位移计发射激光时，从测试室的光线示波器上可直接获取位移图像及有关参数并自动记录成果。

When this observation method is used, the displacement meter is installed on the ground floor or basement of the building, the receiving device can be installed on the top floor or the floor to be observed, the laser channel can be the stairwell, and the test room should be located in the floor near the top. When the displacement meter emits laser, the displacement image and relevant parameters can be directly obtained from the light oscilloscope in the test room and the results can be automatically recorded.

## （2）倾斜观测的实施

## (2) Implementation of inclination observation

①测站点和观测点的布设。

① Layout of observation stations and points.

当从建筑物外部观测时，测站点或工作基点的点位应选在与照准目标中心连线呈接近正交或成等分角的方向线上、距照准目标 1.5~2 倍目标高度的固定位置处。当利用建筑物内竖向通道观测时，可将通道底部中心点作为测站点。

When we observe from the outside of the building, the observation station or working reference point shall be selected at a fixed position 1.5-2 times the height of the target on the direction line nearly orthogonal to or equidistant from the center line of the observation target. When we observe through the vertical channel inside the building, the center point at the bottom of the channel can be used as the observation station.

按纵横轴线或前方交会布设的测站点，每点应选设 1~2 个定向点；基线端点的选设应顾及其测距或丈量的要求。按方向线水平角法布设的测站点，应设置好定向点。

For observation stations arranged on the vertical and horizontal axes or forward intersections, 1-2 orientation points shall be selected for each point; the selection of baseline end points shall take into account the ranging or measurement. Orientation points shall be provided for observation stations arranged using the direction line and horizontal angle method.

观测点应沿对应测站点的某主体竖直线，对整体倾斜按顶部、底部对应布设，对分层倾斜按分层部位、底部上下对应布设。

The observation points shall be arranged along a vertical line of some main body of the corresponding observation station; for overall inclination, the top corresponds to the bottom; for floor inclination, the top corresponds to the bottom of floor bottom.

②倾斜观测点位标志设置。

② Markers setting at inclination observation points.

建筑物顶部和墙体上的观测点标志，可采

用埋入式照准标志形式；有特殊要求时，应专门设计。

Buried sighting marks can be set at the observation points on the top and wall of the building; if there are special requirements, they shall be specially designed.

不便埋设标志的塔形、圆形建筑物以及竖直构件，可以用照准视线所切同高边缘认定的位置或用高度角控制的位置作为观测点位。

For towers, circular buildings and vertical members for which buried marks are not available, the position where the line of sight is tangent to the same height edge or the position controlled by the height angle can be used as the observation point.

位于地面的测站点和定向点，可根据不同的观测要求，采用带有强制对中设备的观测墩或混凝土标石。

Observation pillars or concrete markstones with forced centering equipment can be used for observation stations points and orientation points on the ground according to different observation requirements.

对于一次性倾斜观测项目，观测点标志可采用标记形式或直接利用符合位置与照准要求的建筑物特征部位；测站点可采用小标石或临时性标志。

For one-time inclination observation, marks or characteristic parts of building that meet the position and observation requirements can be used to mark the observation points; small markstones or temporary marks can be used for observation stations points.

③观测周期的确定。

③ Determination of observation period.

主体倾斜观测的周期，可视倾斜速度每1~3个月观测一次。当基础附近因大量堆载或卸载、场地降雨或长期积水等而导致倾斜速度加快时，应及时增加观测次数。

The inclination of the main part of the building should be observed once every 1-3 months depending on the inclination speed. When the inclination speed is accelerated due to a large amount of stacking or unloading near the foundation and long-term rainfall on the site or water accumulation, the observation frequency shall be increased in time.

施工期间的观测周期应符合下列规定。

The observation period during construction shall meet the following requirements.

a. 普通建筑可在基础完工后或地下室砌完后开始观测，大型、高层建筑可在基础垫层或基础底部完成后开始观测。

a. Observation for general buildings can be started after the foundation is completed or the basement is built, and for large and high-rise buildings it can be started after the foundation cushion or foundation bottom is completed.

b. 观测次数与间隔时间应视地基与加荷情况而定。民用高层建筑可每加高1~5层观测一次，工业建筑可按回填基坑、安装柱子和屋架、砌筑墙体、设备安装等不同施工阶段分别进行观测。若建筑施工均匀增高，应至少在增加荷载的25%、50%、75%和100%时各测一次。

b. The observation frequency and intervals shall depend on the foundation and load increase. High-rise civil buildings should be observed every 1-5 additional storeys, and industrial buildings should be observed at different construction stages such as backfilling of foundation pits, installation of pillars and roof trusses, wall construction and equipment installation. If the building construction height is increased evenly, observation shall be carried out at least once respectively when the load is increased by 25%, 50%, 75% and 100%.

c. 施工过程中若暂时停工，在停工时及重新开工时应各观测一次。停工期间可每隔2~3

个月观测一次。

c. If the construction is suspended, observation shall be made at the time of suspension and immediately before the resumption of work. During the downtime, observation frequency may be 2-3 months.

倾斜观测应避开强日照和风荷载影响大的时间段。

The inclination observation shall not be done in the period with strong sunlight and wind load.

④结果提交。

④ Submission of results.

倾斜观测工作结束后，应提交下列观测结果：倾斜观测点位布置图、倾斜观测成果表、主体倾斜曲线图、观测成果分析资料。

After the inclination observation, the following results shall be submitted: layout of inclination observation points, table of inclination observation results, main part inclination curve, analysis data of observation results.

## 7.4 建筑物的水平位移观测

## 7.4 Horizontal Displacement Observation of Building

建筑物水平位移观测包括位于特殊土质地区的建筑物地基基础水平位移观测，受高层建筑基础施工影响的建筑物及工程设施水平位移观测，以及挡土墙、大面积堆载等工程中所需的地基土深层侧向位移观测等，应测定在规定平面位置上随时间变化的位移量和位移速度。根据场地条件，建筑物水平位移观测可采用基准线法、全站仪坐标法等。

The horizontal displacement observation of buildings refers to the horizontal displacement observation of building foundations in special soil areas, the horizontal displacement observation of buildings and construction facilities vulnerable to the construction of high-rise building foundations, and the lateral displacement observation of deep foundation soil required for retaining walls, large-area loading and other works. The displacement and displacement speed changing with time at the specified plane shall be measured. The horizontal displacement may be measured by the reference line method and the total station coordinate method, depending on the site conditions.

基准线法的原理是在与水平位移垂直的方向上建立一个固定不变的铅垂面，测定各观测点相对该铅垂面的距离变化，从而求得水平位移。基准线法适用于直线形建筑物。

The principle of the reference line method is to establish a fixed vertical plane in the direction perpendicular to the horizontal displacement, and measure the distance change of each observation point relative to the vertical plane, so as to obtain the horizontal displacement. The reference line method is applicable to straight buildings.

### (1) 原理

### (1) Principle

建立一条基准线，根据它测定建筑物垂直于基准线方向的水平位移。

First, establish a reference line, and then measure the horizontal displacement of the building in the direction perpendicular to the reference line.

### （2）适用建筑

### (2) Applicable buildings

这种方法适用于变形方向为已知的线形建（构）筑物，如水坝、桥梁等。

It is applicable for linear buildings (structures) with known deformation direction, such as dams and bridges.

### （3）观测方法

### (3) Observation method

基准线法又可分为测小角法和活动觇牌法。

The reference line method can be divided into minor angle method and moving target method.

①测小角法。

① Minor angle method.

测小角法是视准线法的一种，其原理是利用全站仪、经纬仪或准直仪等光学仪器，在两个工作基点之间建立一条工作基线，以该工作基线为依据测定出各个观测点的水平位移量，如图 7-6 所示。

The minor angle method is a kind of collimating line method. Its principle is to establish a working reference line between two working reference points using optical instruments such as total station, theodolite or collimator, and then measure the horizontal displacement of each observation point based on the working reference line, as shown in Figure 7-6.

利用全站仪或经纬仪等光学仪器，在两个基准点之间建立一个基准面，以该基准面为基准测定出各个观测点的水平位移量的方法称为视准线法。

When it comes to the collimating line method, a reference plane is established between two reference points using optical instruments such as total station or theodolite, and then the horizontal displacement of each observation point is measured based on the reference plane.

②活动觇牌法。

② Moving target method.

活动觇牌法原理是首先利用全站仪等光学仪器在两个基准点之间建立一个基准面；然后将活动觇牌置于其中需要测量的一个位移点上，供测量仪器瞄准以测定位移点偏离值，在觇牌的读数设备上读取读数（可估读到 0.1 mm）。因此要求活动觇牌上有分划尺，最小分划值为 1 mm，用游标卡尺可直接读到 0.1~0.01 mm。使用该方法观测简单、清晰，如图 7-7 所示。

The principle of the moving target method is to establish a reference plane between two

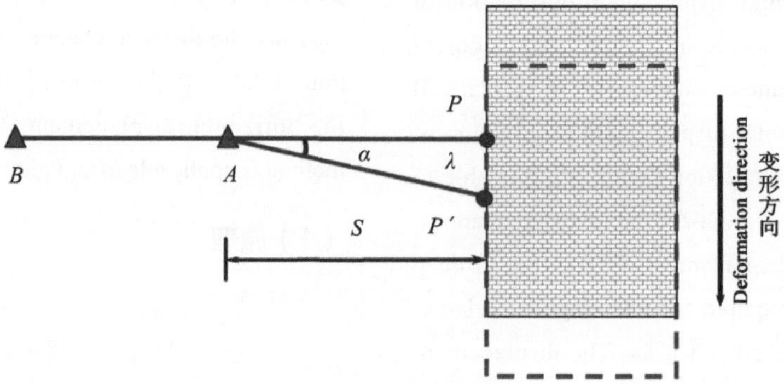


图7-6　测小角法

Figure 7-6 Minor Angle Method

reference points using optical instruments such as a total station; then, the moving target is placed on a displacement point to be measured, and the measuring instrument aims at it to measure the displacement point deviation value, and the reading is taken on the target reading device (which can be read to 0.1 mm). Therefore, scale values are required on the moving target, and the minimum scale value is 1mm, which can be directly read to 0.1-0.01 mm with a vernier caliper. The method is simple and clear, as shown in Figure 7-7.

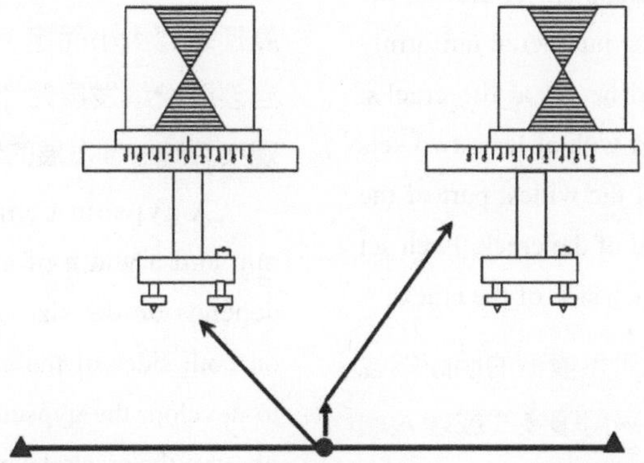


图7-7　活动觇牌法

Figure 7-7 Moving Target Method

# 7.5　建筑物的裂缝观测

# 7.5 Observation of Building Cracks

## 7.5.1 裂缝观测的内容

## 7.5.1 Contents of Crack Observation

裂缝观测的内容是测定建筑物上的裂缝分布位置，裂缝的走向、长度、宽度及变化程度。观测的裂缝数量应视需要而定，对主要的或变化大的裂缝应进行观测。

Crack observation is to measure the distribution position, direction, length and width of cracks on buildings. The number of cracks to be observed shall be determined according to the needs, and the major cracks or cracks with large changes shall be observed.

## 7.5.2 裂缝观测点的布设和观测标志

## 7.5.2 Layout of Crack Observation Points and Observation Marks

发现建筑物有裂缝，除了要增加沉降观测的次数外，还应立即进行裂缝变化的观测。为了观测裂缝的发展情况，首先要对需要观测的裂缝进行统一编号，并在裂缝处设置观测标志。

每条裂缝至少应布设两组观测标志：一组在裂缝最宽处，另一组在裂缝末端。每组标志由裂缝两侧各一个标志组成。

When cracks are found in a building, the frequency of settlement observation shall be increased and crack change observed immediately. In order to observe the development of cracks, the cracks to be observed shall be numbered uniformly and observation marks shall be set at the cracks. Each crack shall be provided with at least two sets of observation marks: one at the widest part of the crack and the other at the end of the crack. Each set of marks consist of one on each side of the crack.

裂缝观测标志应具有可供量测的明晰端或中心。观测期较长时，可采用镶嵌式或埋入墙面的金属片标志、金属棒标志或楔形板标志；观测期较短或要求不高时，可采用油漆平行线标志或用建筑胶粘贴的金属片标志；要求较高、需要测出裂缝纵横向变化值时，可采用坐标方格网板标志；使用专用仪器设备观测的标志，可按具体要求另行设计。

Crack observation marks shall have clear ends or centers for measurement. When there is a long observation period, metal sheets marks, metal bars marks signboards, metal bars or wedge shaped plates marks can be inlaid or embedded in a wall; when there is a short observation period or low requirement, paint parallel lines marks or metal sheets marks pasted with construction adhesive can be used; when there is a high requirement and the vertical and horizontal changes of cracks need to be measured, the coordinate square grid plates marks can be used; for observation with special instruments and equipment, marks can be designed separately according to specific requirements.

设置标志的基本要求是，当裂缝发展时标志能相应地开裂或变化，以正确地反映建筑物裂缝的发展情况。

The basic requirement for setting up the marks is that the marks can crack or change accordingly when the crack is developed, so as to correctly reflect the development of the building crack.

### （1）石膏板标志

(1) Gypsum marks

用厚 10 mm、宽约 50~80 mm 的石膏板（长度视裂缝大小而定），在裂缝两边固定牢固。当裂缝继续发展时，石膏板也随之开裂，从而观察裂缝继续发展的情况。

A gypsum board with a thickness of 10 mm and a width of about 50-80 mm (the length depends on the size of the crack) is fixed firmly on both sides of the crack. As the crack continues to develop, the gypsum board also cracks, so as to observe the crack development.

### （2）白铁板标志

(2) Galvanized iron plates marks

如图 7-8 所示，用两块白铁板，一块取 150 mm × 150 mm 的正方形，固定在裂缝的一侧，并使其一边和裂缝的边缘对齐。另一块为 50 mm × 200 mm 的长条形，固定在裂缝的另一侧，并使其中一部分紧贴相邻的正方形白铁板。当两块白铁板固定好以后，在其表面均涂上红色油漆。如果裂缝继续发展，两白铁板将逐渐拉开，露出正方形白铁板上原被覆盖没有涂油漆的部分，其宽度即为裂缝加大的宽度，可用尺子量出。

As shown in Figure 7-8, one 150 mm×150 mm galvanized iron plate is fixed on one side of the crack, with one side aligned with the edge of the crack. The other one 50 mm×200 mm (strip) is fixed on the other side of the crack, and a part of it is closely attached to the adjacent square iron plate. When the two pieces of galvanized iron plates are fixed properly, their surfaces are painted red. If the crack continues to develop, the two iron plates will gradually pull apart to expose on the square iron

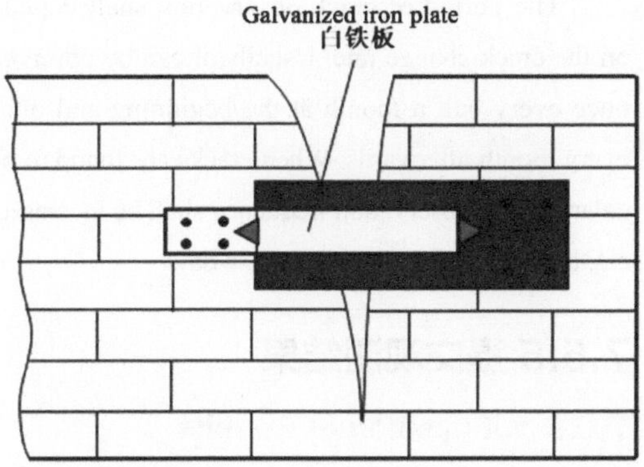


图7-8 白铁板标志

Figure 7-8 Galvanized Iron Sheets Marks

plate the covered part with no paint, its width is the width of the increased crack and can be measured with a ruler.

### （3）金属棒标志

### (3) Metal bars marks

如图 7-9 所示，在裂缝两边凿孔，将长约 10 cm、直径 10 mm 以上的钢筋头插入（在两钢筋头埋设前，应先把钢筋一端锉平，在上面刻画十字线或中心点，作为量取其间距的依据），并使其露出墙外约 2 cm 左右，用水泥砂浆填灌牢固。待水泥砂浆凝固后，量出两金属棒之间的距离，并记录下来。以后如裂缝继续发展，金属棒的间距也就会不断加大，定期测量两棒的间距并进行比较，即可掌握裂缝发展情况。

As shown in Figure 7-9, holes are drilled on both sides of the crack, in which the steel bar heads with a length of about 10 cm and a diameter of more than 10 mm are inserted (before the two steel bar heads are buried, one end of the steel bar shall be filed flat, and a cross line or center point shall be marked on it as a reference for measuring the spacing between the bars) and left exposed by about 2 cm; next, joints are filled firmly with cement mortar. After the cement mortar has set, the distance between the two metal bars are measured and recorded. If the crack continues to develop in

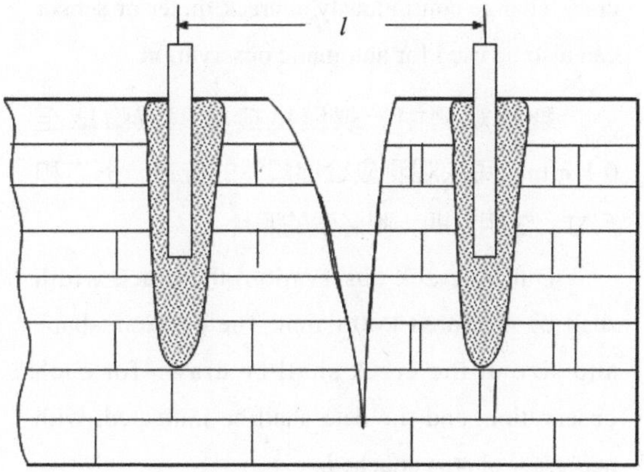


图7-9 金属棒标志

Figure 7-9 Metal Bars Marks

the future, the spacing between the metal bars will continue to increase. The spacing between the two bars should be measured and compared regularly to learn the development of the crack.

## 7.5.3 裂缝观测的方法

## 7.5.3 Crack Observation Method

对于数量不多、易于量测的裂缝，可视标志形式不同，用比例尺、小钢尺或游标卡尺等工具定期量出标志间距离，以求得裂缝变位值，或用方格网板定期读取坐标差以计算裂缝变化值；对于较大面积且不便于人工量测的众多裂缝宜采用近景摄影测量方法；当需连续监测裂缝变化时，还可采用裂缝计或传感器自动测记方法观测。

For a small number of cracks that are easy to be measured, the distance between monuments may be measured regularly with scale, small steel ruler or vernier caliper, etc. (depending on the selection of monuments), so as to obtain the crack displacement value; the coordinate difference shall also be read regularly with a square grid plate to calculate the crack change value; the close-range photogrammetry should be adopted for many cracks with large area and difficulty for manual measurement; when it is necessary to monitor the crack change continuously, a crack meter or sensor can also be used for automatic observation.

裂缝观测中，裂缝宽度数据应量取至0.1 mm，每次观测应绘出裂缝的位置、形态和尺寸，注明日期，附必要的照片。

During crack observation, the crack width shall be measured to 0.1 mm. The position, shape and size of the crack shall be drawn for each observation, and the date shall be indicated, with necessary photos attached.

### 7.5.4 裂缝观测的周期

### 7.5.4 Crack Observation Period

裂缝观测的周期应视裂缝变化速度而定。通常情况下，观测前期可半月测一次，后期一月左右测一次。当发现裂缝加大时，应增加观测次数，可以几天一次或每天一次连续观测。

The period of crack observation shall depend on the crack change rate. Usually, it can be observed once every half a month at the beginning and once every month afterwards. When cracks are found to be enlarged, the observation frequency shall be increased, e.g. once every few days or once a day.

### 7.5.5 提交观测结果

### 7.5.5 Submission of Results

观测工作结束后，应提交下列观测结果：裂缝分布位置图、裂缝观测结果表、观测结果分析说明资料。

After the observation, the following results shall be submitted: crack distribution map; table of crack observation results; analysis and explanation data of observation results.

## 7.6 建筑物的竣工测量

## 7.6 Acceptance Survey of Building

### 7.6.1 竣工测量

### 7.6.1 Acceptance Survey

建(构)筑物竣工验收时进行的测量工作，称为竣工测量。为做好竣工总平面图的编制工作，应随着工程施工进度，同步记载施工资料，并根据实际情况，在竣工时，进行竣工测量。竣工测量主要是对施工过程中设计有更改的部分，进行测量前直接在现场指定施工过程中设计有更改的部分及资料不完整、无法查对的部分，根据施工控制网进行现场实测或加以补测。对于有下列情况之一者或对于下列对象，必须进行现场实测。

The acceptance survey involves the survey work carried out at the completion acceptance of buildings (structures). In order to prepare the general as-built plan, the construction data shall be recorded synchronously with the project construction progress, and the acceptance survey shall be carried out after completion based on the actual situation. The acceptance survey is mainly for the parts with design changes in the construction process, the observers designate directly the parts with design changes in the construction process and the parts that cannot be checked due to incomplete data; the field survey or supplementary survey shall be done according to the construction control network. The field survey must be done in one of the following cases or to the following objects.

①不能及时提供建筑物或构筑物的设计坐标，而在现场制定施工位置的工程。

① The design coordinates of the building or structure are not available in time, so the construction location are determined on site.

②设计图上只标明地物相对尺寸以致无法推算坐标和标高的地物点。

② The design drawings only indicate the relative dimensions of the ground objects, so that the coordinates and elevations cannot be calculated.

③由于设计多次变更而无法查找到确定的设计资料的。

③ The design data confirmed cannot be found due to multiple design changes.

④竣工现场的竖向布置、围墙和绿化情况，施工后尚保留的大型临时设施。

④ Vertical layout, enclosures and greening on the completion site, and large temporary facilities that remain after construction.

### （1）竣工测量的内容

### (1) Work contents of acceptance survey

建筑工程竣工测量内容主要包括以下几个方面。

The acceptance survey of building mainly includes the following.

①一般工业与民用建筑。测量房屋角点坐标及高程，对较大的矩形建筑物至少要测三个主要房角坐标，小型房屋可测其长边两个房屋角点，并量出房宽注于图上，还应测量各种管线进出口位置和高程。

① General industrial and civil buildings. The coordinates and elevations of the corner points of buildings shall be measured. For large rectangular buildings, at least three main corner points shall be measured. For small buildings, the corner points on their long sides may be measured, and the building width shall be measured and indicated on the drawing; the positions and elevations of various pipeline inlets and outlets shall be measured.

②铁路和公路。测量线路的起始点、转折点、曲线起始点、曲线元素、交叉点坐标、桥涵等构筑物位置和高程。

② Railways and highways. The starting points, turning points, curve starting points, curve elements, coordinates of intersection points of lines, and locations and elevations of bridges, culverts and other structures shall be measured.

③地下管线。测量管线转折点、起点及终点的坐标，测量、检查井旁地面、井盖、井底、沟槽、井内敷设物和管顶等处的标高。

③ Underground pipelines. The coordinates of the turning point, beginning point and ending point of the pipeline shall be measured, and the elevations of the ground, well cover, well bottom, trench, objects laid in the well and pipe top, etc. shall be measured and checked.

④架空管线。测量管线转折点、结点、交叉点的坐标，测量支架间距及支架旁地面标高、基础标高，管座、最高和最低电线至地面的净高等。

④ Overhead pipeline. The coordinates of pipeline turning point, junction and intersection, the spacing between supports, the ground elevation beside supports, the foundation elevation, the clear heights of pipe seats and the highest and lowest wires to the ground, etc. shall be measured.

⑤特种构筑物。测量沉淀池、烟囱、煤气罐等及其附属构筑物的外形和四角的坐标，圆形构筑物的中心坐标，基础标高，构筑物高度，沉淀池深度等。

⑤ Special structures. The shapes and coordinates of four corners of the sedimentation tank, chimney, gas tank, ancillary structures, etc., the center coordinates of circular structures, the foundation elevation, the elevation of structure, the depth of sedimentation tank, etc. shall be measured.

⑥其他。测量围墙拐角点坐标、绿化区边界以及一些不同专业需要反映的设施和内容。

⑥ Others. The coordinates of corner points of the enclosures, the boundary of the green area and some facilities and contents that need to be reflected by different disciplines shall be measured.

### (2)竣工测量方法的特点

### (2) Characteristics of acceptance survey method

竣工测量的基本测量方法与地形测量相似，区别在于以下几点。

The acceptance survey is basically similar to the topographic survey, with the following differences.

①一般竣工测量图根控制点的密度要大于地形测量图根控制点的密度。

① Generally, the acceptance survey mapping control points are denser than topographic survey mapping control points.

②地形测量一般采用视距测量的方法测定碎部点的平面位置和高程，而竣工测量一般采用经纬仪测角、钢尺量距的极坐标法测定碎部点的平面位置，采用水准仪或经纬仪测定碎部点的高程，也可用全站仪进行测绘。

② For topographic survey, the plane positions and elevations of detail points are generally determined by stadia survey, while for acceptance survey the plane positions of detail points are generally determined using the polar coordinate method, such as angle measurement with theodolite and distance measurement with steel tape, and the elevations of detail points are determined by level or theodolite, or by total station.

③竣工测量的测量精度要高于地形测量的测量精度。地形测量的测量精度要求满足图解精度，而竣工测量的测量精度一般要满足解析精度，应精确至厘米。

③ The accuracy of acceptance survey is higher than that of topographic survey. The accuracy of topographic survey is required to meet the graphic accuracy, while the accuracy of acceptance survey shall generally meet the analytical accuracy, which shall be accurate to centimeter.

④竣工测量的内容比地形测量的内容更丰富。竣工测量不仅测地面的地物和地貌，还要测地下各种隐蔽工程，如上水管线、下水管线及热力管线等。

④ The acceptance survey has more contents than topographic survey. The work contents of acceptance survey include not only the ground objects and landform, but also various underground concealed works such as water supply pipelines, sewage pipelines and thermal pipelines.

## 7.6.2 竣工总平面图的编绘

## 7.6.2 Compilation of General As-built Plan

建设工程项目竣工后，应编绘竣工总平面图。竣工总平面图是设计总平面图在施工后实际情况的全面反映，工业与民用建筑工程是根据设计总平面图施工的。在施工过程中，种种原因都有可能使建(构)筑物竣工后的位置与原设计位置不完全一致，所以，设计总平面图不能完全代替竣工总平面图。

After the completion of the construction project, the general as-built plan shall be prepared. The general as-built plan is a comprehensive reflection of the actual situation after construction according to the design general plan, and the industrial and civil building works are constructed according to the design general plan. In the construction process, due to various reasons, the position of the completed building (structure) is not completely consistent with the initial design position, so the general design plan cannot completely replace the general as-built plan.

### （1）编绘竣工总平面图的目的

### (1) Purpose of compiling the general as-built plan

①反映设计的变更情况。施工过程中由于发生设计时未考虑到的问题而要变更设计，这种临时变更设计的情况必须通过测量反映到竣工总平面图上。

① Reflection of design changes. If the design is changed due to problems not considered in the design during the construction process, such temporary design change must be reflected on the general as-built plan through measurement.

②提供各种设备的维修依据。竣工总平面图可以为各种设备、设施进行维修工作时提供数据。

② Provision of maintenance basis for various equipment. The general as-built plan can provide data for maintenance of various equipment and facilities.

③保存建筑物的历史资料。竣工总平面图可以提供原有建筑物、构筑物、地下和地上各种管线和交通路线的坐标及坐标系统、高程及高程系统等重要的历史资料。

③ Preservation of historical data of buildings. The general as-built plan can provide important historical data such as coordinates and coordinate system, elevation and elevation system of existing buildings, structures, underground and above-ground pipelines and traffic routes.

### （2）竣工总平面图的内容

### (2) Contents of general as-built plan

竣工总平面图应包括控制点，如建筑方格网控制桩点位、水准点、建筑物平面位置、辅助设施、生活福利设施、架空与地下管线，还应包括铁路等建筑物或构筑物的平面施工放线坐标、高程以及室内外平面图。竣工总平面图一般采用 1 ：1 000 比例尺绘制，若要清楚表示局部地区，也可采用 1 ：500 比例尺绘制。

The general as-built plan shall include control points, such as control peg points of building square grid, benchmarks, building plane positions, auxiliary facilities, living welfare facilities, overhead and underground pipelines, as well as plane construction setting-out coordinates, elevations and indoor and outdoor plans of buildings or structures such as railways. The general as-built plan is generally drawn at a scale of 1 ：1 000, or at a scale of 1 ：500 if local areas are to be clearly indicated.

### （3）竣工总平面图的编绘步骤

### (3) Compilation steps of general as-built plan

①确定竣工总平面图的比例尺。竣工总平面图的比例尺，应根据企业的规模大小和工程的密集程度参考下列规定。

① Determine the scale of the general as-built plan. The scale of the general as-built plan shall consider the following, depending on the size of the enterprise and the intensity of the project.

a. 小区内为 1 ：500 或 1 ：1 000。

a. 1 ：500 or 1 ：1 000 in the plot.

b. 小区外为 1 ：1 000~1 ：5 000。

b. 1 ：1 000~1 ：5 000 outside the plot.

②绘制竣工总平面图图底坐标方格网。为了能长期保存竣工资料，竣工总平面图应采用质量较好的图纸。聚酯薄膜具有坚韧、透明、不易变形等特性，可用作图纸。

② Draw the bottom coordinate square grid of the general as-built plan. In order to keep the as-built drawing for a long time, the paper of good quality shall be used. The polyester film is tough, transparent and not easy to deform, so it can be used.

③展绘控制点。以图底上绘出的坐标方格网为依据，将施工控制网点按坐标展绘在图上。展点对所邻近的方格而言，其允许偏差为±0.3 mm。

③ Plotting of control points. Based on the coordinate square grid on the bottom of the drawing, the construction control network points shall be plotted on the drawing according to the coordinates. The allowable deviation of the point relative to the adjacent square is 0.3 mm.

④展绘设计总平面图。在编绘竣工总平面图之前，应根据坐标方格网，先将设计总平面图的图面内容按其设计坐标，用铅笔展绘于图纸上，作为底图。

④ Plotting of design general plan. Before compilation of the general as-built plan, the contents of the design general plan shall be plotted on the drawing with a pencil based on its design coordinates and the coordinate grid as the base drawing.

⑤展绘竣工总平面图。对按设计坐标进行定位的工程，应以测量定位资料为依据，按设计坐标（或相对尺寸）和标高展绘。对原设计进行变更的工程，应根据设计变更资料展绘。对有竣工测量资料的工程，若竣工测量成果与设计值之差不超过所规定的定位容许误差时，按设计值展绘；否则，按竣工测量资料展绘。竣工总平面图编绘完成后，应经原设计及施工单位技术负责人审核、会签。

⑤ Plotting of general as-built plan. For projects positioned according to the design coordinates, the survey and positioning data shall be taken for reference, and the plan shall be drawn according to the design coordinates (or relative dimensions) and elevation. For projects with design changes, the plan shall be drawn according to the design change data. For projects with acceptance survey data, if the ratio difference between the acceptance survey result and the design value does not exceed the specified allowable positioning error, it shall be drawn according to the design value; otherwise, it shall be drawn according to the acceptance survey data. After the completion of the general as-built plan, it shall be reviewed and countersigned by the technical directors of the original designer or the construction contractor.

## 7.6.3 竣工总平面图的绘制

## 7.6.3 Other Precautions and Contents of General As-built Plan

### （1）分类编绘竣工总平面图

### (1) Compiler as-built plan categorically

对于大型企业和较复杂的工程，如将厂区地上、地下所有建筑物和构筑物都绘制在一张总平面图上，则会导致图面线条密集，不易辨认。为了使图面清晰醒目，便于使用，可根据工程的密集与复杂程度，按工程性质分类编绘竣工总平面图。

For large enterprises and more complex projects, if all underground and above-ground buildings and structures in the plant area are drawn on a general plan, there will be dense lines on the plan and it is difficult to recognize. In order to make the drawing clear and legible, the general as-built plan can be compiled based on the density and complexity of the project and the nature of the project.

### （2）综合竣工总平面图

### (2) Comprehensive general as-built plan

综合竣工总平面图即全厂性的总体竣工总平面图，包括地上、地下一切建筑物、构筑物和竖向布置及绿化情况等。

The comprehensive general as-built plan is the overall general as-built plan of the whole plant, including all underground and above-ground buildings, structures, vertical layout and greening.

### （3）竣工总平面图的图面内容和图例

### (3) Contents and legend of general as-built plan

竣工总平面图的图面内容和图例，一般应与设计图一致。图例不足时可补充编绘。

The contents and legend of the general as-built plan shall generally be consistent with the design drawings. If the legend is insufficient, it can be supplemented.

### （4）竣工总平面图的附件

### (4) Attachments of general as-built plan

为了全面反映竣工成果，便于生产、管理、维修和日后企业的扩建或改建，与竣工总平面图有关的一切资料，应分类装订成册，作为竣工总平面图的附件保存。

In order to fully reflect the completion results and facilitate production, management, maintenance and future expansion or reconstruction of the enterprise, all materials related to the general as-built plan shall be sorted and bound in a volume and kept as an attachment to the general as-built plan.

### （5）工业企业竣工总平面图

### (5) General as-built plan of industrial enterprises

对于工业企业竣工总平面图的编绘，最好的办法是随着单位或系统工程的竣工，及时地编绘单位工程或系统工程平面图，并由专人汇总各单位工程平面图，进而编绘竣工总平面图。

It's best for an industrial enterprise to prepare the plans of unit works or system works in time with the completion of unit works or system works, and assign special personnel to gather the plans of all unit works to compile the general as-built plan.

# 本章小结

# Chapter Summary

本章主要讲述了建筑工程变形观测的基本知识、建筑物的沉降观测、建筑物的倾斜观测与裂缝观测、建筑物的水平位移观测和建筑物的竣工测量等内容。

This chapter mainly describes the basic knowledge of building deformation observation, building settlement observation, building inclination observation, building crack observation, building horizontal displacement observation, and acceptance survey.

1. 变形观测的任务是周期性地对建筑物上的观测点进行重复观测，以求得观测点位置的变化量。

1. The task of deformation observation is to repeatedly observe the observation points on the building periodically in order to obtain the variation quantity of the observation points.

2. 沉降观测常采用水准测量的方法，根据布设好的水准基点和沉降观测点，布设闭合或者附合水准路线，采用水准路线计算方法对高差闭合差进行分配，从而对计算结果进行检校，最终计算出水准路线上的每个观测点的高程。

2. Leveling is often adopted for settlement observation. According to the benchmarks and settlement observation points, closed or annexed leveling lines are arranged, and the elevation difference closure error is distributed using the leveling line calculation method, so as to check the calculation results and finally calculate the elevation of each observation point on the leveling line.

3. 测定建筑物倾斜度随时间而变化的工作叫作倾斜观测。倾斜观测一般用水准仪、经纬仪、垂球或其他专用仪器来测量建筑物的倾斜度。

3. The inclination observation involves measuring the inclination of building with time. Generally, we measure the building inclination with a level, theodolite, plumb or other special instruments.

4. 当建筑物产生裂缝时，应立即进行观测，了解其现状并掌握其发展情况，并根据观测所得到的资料分析裂缝产生的原因和它对建筑物安全的影响程度，及时采取有效措施加以处理。

4. When a crack occurs in a building, it shall be observed immediately to understand its current situation and development, and the cause of the crack and the impact on the safety of the building shall be analyzed based on the data obtained from the observation, and effective measures shall be taken immediately to deal with it.

5. 水平位移观测可使用视准线、激光准直、测边角等方法。

5. The horizontal displacement can be observed using the collimation line method, laser collimator method, corner measurement method, etc.

6. 建（构）筑物竣工验收时进行的测量工作，称为竣工测量。为做好竣工总平面图的编制工作，应随着工程施工进度，同步记载施工资料，并根据实际情况，在竣工时进行竣工测量。

6. The acceptance survey involves the survey work carried out at the completion acceptance of buildings (structures). In order to compiler the general as-built plan, the construction data shall be recorded synchronously with the project construction progress, and the acceptance survey shall be carried out after completion based on the actual situation.

7. 建设工程项目竣工后，应编绘竣工总平面图。竣工总平面图是设计总平面图在施工后实际情况的全面反映，工业建筑工程与民用建筑工程是根据设计总平面图来进行施工的。

7. After the completion of the construction project, the general as-built plan shall be compiled. The general as-built plan is a comprehensive reflection of the actual situation after construction according to the design general plan, and the industrial and civil building works are constructed according to the design general plan.

# 复习思考题
# Review Questions

## 一、填空题
## I. Fill in the Blanks

1. 对于特等、一等沉降观测，基准点不应少于______个；对于其他等级沉降观测，基准点不应少于______个。

1. For special and first order settlement observation, there shall be no less than______benchmarks, and no less than ______ benchmarks for settlement observation of other classifications.

2. 当半年内沉降量不超过______mm时，便认为沉降趋于稳定。

2. When the settlement does not exceed ______mm within half a year, it is considered to be stable.

3. 倾斜观测一般是用______、______、______或其他专用仪器来测量建筑物的倾斜度。

3. Generally, we measure the building inclination with a______, ______, ______or other special instruments.

4. 对需要观测的裂缝应统一进行编号，每条裂缝至少应布设______组观测标志。

4. The cracks to be observed shall be numbered uniformly, and ______ observation marks shall be set for each crack at least.

5. 建筑水平位移观测点的位置应选在______、______、______等处。

5. The horizontal displacement observation points of the building shall be selected at______, ______, ______, etc.

## 二、选择题
## II. Choice Questions

1. 无论何种建筑物，沉降观测次数不能少于（ ）次。

1. No matter what kind of building, the settlement observation shall be done for no less than ( ) times.

A.3　　B.4　　C. 5　　D. 6

2. 在民用建筑中，通常在（ ）部位设置观测量。

2. In civil buildings, observation is usually set at ( ).

A. 房屋的转角　　B. 沉降缝两侧

A. corners of buildings　　B. both sides of settlement joint

C. 基础的对称轴上

C. symmetry axis of foundation

D. 地质条件改变处

D. a place with change in geological conditions

3. 观测期较短时，裂缝观测标志可采用（ ）。

3. ( ) can be used for crack observation monuments when the observation period is short.

A. 镶嵌或埋入墙面的金属标志

A. Metal marks inlaid or embedded in walls

B. 金属棒标志

B. Metal bars marks

C. 油漆平行线标志

C. Paint parallel lines marks

D. 坐标方格网板标志

D. Coordinate square grid plates marks

4. 当测量地面观测点在特定方向的位移时，可使用（ ）方法。

4. When measuring the displacement of a ground observation point in a particular direction, the ( ) method can be used.

A. 前方交会

A. forward intersection

B. 视准线

B. collimation line

C. 激光准直

C. laser collimation

D. 测边角

D. corner measurement

## 三、简答题

## III. Short Answer Questions

1. 建筑物为什么要进行变形观测？

1. Why is the building deformation observation required?

2. 水准基点的布设应符合哪些要求？

2. What are the requirements for the layout of benchmarks?

3. 简述沉降观测中常遇到的问题及其处理方法。

3. Briefly describe the common problems in settlement observation and solutions.

4. 简述裂缝观测的方法。

4. Briefly describe how to carry out crack observation.